INSECT MUSEUM

Published by Firefly Books Ltd. 2009

First printing

Publisher Cataloging-in-Publication Data (U.S.)

Dourlot, Sonia.
Insect museum / Sonia Dourlot.
Originally published as: Petite Collections d'Insects ; Larouse: France, 2008.
[256] p. : col. photos. ; cm.
Includes bibliographical references and index.
Summary: Examines over 110 different insects, including their characteristics, distribution and size.
ISBN-13: 978-1-55407-483-9
ISBN-10: 1-55407-483-5

1. Insects. 2. Insects – Pictorial works. II. Title.
595.7 dc22 QL467.D687 2009

Library and Archives Canada Cataloguing in Publication

Dourlot, Sonia
Insect museum / Sonia Dourlot.
Translation of Petite collection d'insectes de nos régions.
Includes bibliographical references and index.
ISBN-13: 978-1-55407-483-9
ISBN-10: 1-55407-483-5
1. Insects. 2. Insects–Pictorial works. I. Title.
QL467.D6813 2009 595.7 C2009-901760-1

Published in the United States by
Firefly Books (U.S.) Inc.
P.O. Box 1338, Ellicott Station
Buffalo, New York 14205

Published in Canada by
Firefly Books Ltd.
66 Leek Crescent
Richmond Hill, Ontario L4B 1H1

Printed in Spain

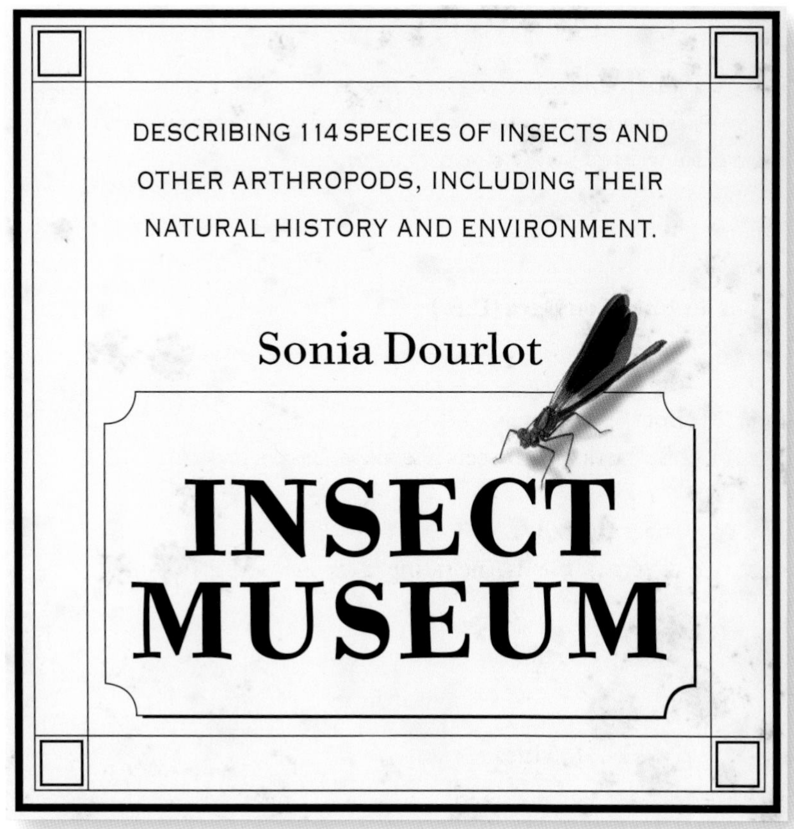

FIREFLY BOOKS

Author's Prologue

Not so harmful "tiny beasts"

Whether we wish to admit it or not, these "little beasts" share our daily routines. They are plentiful, widespread and generally given a bad reputation. However, ignorance is often at the core of the fears surrounding them. Each year, with the coming of spring, we witness the reappearance of these now-familiar insects. However, we may know very little about their behavior and the ecological role they play in nature. The ones we refer to most frequently are, of course, those that intimidate us the most. Among these are some hymenopteras (wasps and hornets), which sting as a defensive response; female hematophages (mosquitoes, horseflies and ticks), which need blood proteins to guarantee the maturation of their eggs; as well as garden destroyers (potato beetles, bark beetle larvae and caterpillars) and apartment squatters (cockroaches and spiders). That's about the extent of it. All of these undesirable species are few compared to countless others. Basically, all play some role in maintaining the equilibrium of our ecosystem.

Often out of self-interest, we have become disrespectful of nature even though we wouldn't exist without it. Critical to our survival, nature gives us water, earth and the very air that we breathe. From this very nature, arthropods represent the largest number of species of all living organisms. What would become of our planet without them? Who would pollinate the flowers to produce the fruits and vegetables essential to our survival? Who would contribute to the decomposition of detritus, waste materials and cadavers? Who would fertilize our soils? Who would control the proliferation of certain plants? What would birds eat without them? Insects occupy a determinant place in all ecosystems. They make up an essential link in the food chain by regulating the populations of numerous species of plants and animals.

Let's end the prejudice

It is not necessarily the fault of their own that arthropods elicit disgust or cause fear. These feelings are mostly generated by adults who transmit them on to their children. Frightening stories, often lacking in actual fact, are passed on from one generation to the next. Instead, we should be encouraging our children to discover the fascinating world of these tiny beasts.

Best to observe them

To better know insects, we don't need to pierce them with a fine pin and line them up in a display case. This is not to say that this method is not useful when it's necessary to study an arthropod family or an environment. A collection then becomes a reference tool that greatly facilitates their identification. Entomologists collect insects to identify them and to create a record of the species, with the goal of protecting them – often a misunderstood motivation. The impact of these collections is small compared to the repercussions resulting from applying insecticides, destroying a biotope or even the act of mowing a lawn! This is why I always encourage children who wish to start a collection. They will learn a great deal and, if they persevere in their pursuit, could later contribute to the inventory of regional species.

Whatever the outcome, nothing is more fascinating than the direct observation of living insects. While reading this book, you will encounter advice on how to attract them to best study them, and certain precautions to take in safely handling a small minority of them. Even if the capture of animals is something I disapprove in principle, I can't but appreciate the sight of a curious child capturing an insect to observe it.

Continuing out traditional passions

I grew up in the country in a verdant region which is very dear to me – the Franche-Comté in eastern France. My friends and I spent all our time in this generous natural environment. And possibly because it was in my temperament, I became a naturalist. I still remember what was undoubtedly my first entomological experience on a particular night at my grandparents'. I fell asleep while staring at the tiny glow in a glass jar that came from a firefly I had captured, which I had placed, with a bit of soil, in a transparent box on my night table. Later on, I became engrossed while reading *Souvenirs Entemologiques* by Jean Henri Fabre and, and all of his subsequent writings. I avidly share in his moments of discovery. I became eager to observe what he had described so well and to create images that would demonstrate, in my own manner, the beauty and complexity of these insects. To develop the text for this book, I delighted in immersing myself in the writings of de Béaumur, the natural history works of Blanchard and Figuier, and also Brehm. I cannot forget *La Hulotte*, for which I impatiently awaited each new installment. The film *Microcosmos*, directed by Claude Nuridsany and Marie Pérennou, achieved well-known success. Their images, which have stayed with me to this day, have finally allowed the public at large to discover insects as they are in reality, in all their beauty and magic, soaring above clichés, prejudices and ignorance.

In 1999, as I was leaving advertising photography, with no specific career path in mind, a position as a photographer in an entomological research laboratory at the University of Rennes became available. This finally provided me with the opportunity to simultaneously satisfy both of my passions. To this day, to finalize a shoot, I display specimens that I am still in awe of. The extraordinary energy displayed by these little animals, the abilities they develop, the morphological adaptations they display and the ingenious strategies they employ to guarantee their survival both surprise and fascinate me more and more each and every day. Undoubtedly very influenced by certain entomological illustrators such as Bernard Durin, and by my advertising experience, I quickly developed an urge to photograph the insects in their environment, just posed on a neutral background, to better appreciate the details.

This book is now yours to enjoy ...

This work has been undertaken to satisfy the curiosity of those who wish to know more about the natural history of the little beasts that live among us. Another goal is to closer examine them with the aid of photography so as to discover details impossible to see at their actual size. I have selected 114 species among the hundreds of thousands, based on criteria of popularity and for the astonishment they might elicit. We need not visit the tropics to become amazed, since most of the species described herein are found close to our homes. It has been my hope, through this work, to pay homage to those creatures that guarantee the stability of our ecosystems. Have fun perusing here and there all the interesting information provided in this book. I sincerely hope that reading it will allow you to fully appreciate the true value of the little animals that populate our gardens and live in our houses. To those curious of nature who can still be in awe and who take the time to observe, have a great time!

Sonia Dourlot

ACKNOWLEDGMENTS

To put the finishing touches on a book, you need an editor. Therefore, my thanks to Editions Larousse for the confidence they expressed in me. Also, you must surround yourself with helpful people. My sincerest gratitude to Marléne Goubalt Body, my little sister, for your unconditional availability, your innumerable corrections, your encouragement and your wonderful friendship.

To my colleages at the University of Rennes, with whom I share my years of work–Denis Poinsot, Sébastien Dugravot, Anne Lizéet and, in particular, Anne Marie Cortesero–please accept my wholehearted gratitude for the high-fives, for the scientific advice, for your support and for your friendship.

To my friend and talented photographer Bernard Chaubet, who supplied certain missing images, I tip my hat.

Thanks also to Patrice Leraut and to Thierry Olivaux for their contribution to this manuscript.

In their respective areas, for the many critical rereads and for their constructive criticism, I wish to extend my gratitude to: Christine Rollard, Jean-Jacques Geoffroy, Étienne Lorio, Yean-Yes Robert, Bruno Michel, Jean-Pierre Néson, Jean-Claude Stretto, Jean-Claude Martin, Mathurin Body, Claudine Pérez-Eid, Raymond Borneck, Jaqueline and Jean-Sébastien Pierre, Julien Petillon, Johanna Villenave, Cathy Caldumide, Jean-Pierre Lumanet, Olivier Plantard and Nico Schneider. Forgive me if I have forgotten anyone.

To my valued researchers, producers, identifiers or collectors of insects: Vincent Harraca, Claire Villemant, Gabriel Haguet, Xavier Gouverneur, Claude Thomé, Emmanuel Delposse, Mathieu Giacomino, Roselyne Souriau (Saveol), Christian Hécker (Iftech), Jean-Pierre Richard, Émilien Lasne, Fabienne Blanschon (SRVP), Shenna Cotter, Jean Kress and Paul Amiet. Thanks to each of you for your contribution.

To Bernard Le Garf who resurrects dead languages, for your friendship I thank you muchly.

To Alain Canard and Jean-Michel Lemaire, for your generous knowledge I bestow my gratitude.

To Muriel Chevier, for your help and conviviality in the midst of Cretia, thanks again.

For generous logical help and computer assistance, a big thanks to Xavier Marquer.

To the numerous people who, in their own way, contributed their individual stones to the construction of this building: my mother, Nicole Dourlot, and to Jean-Yves Gerbet, who propped me up in times of fatigue; to Sylvie Laroche, Évelyne Fèvre-Burdy, for many years of support; to Fabienne Fèvre-Burdy for your companionship during my entomological hunting on our cherished Jurassic Park; to Hervé Brugnot for all the years of wrestling with the countryside, and to Michel, your father, who had the great idea of digging into the mud to make it a paradise for insects; to Alexandre François for our discoveries, but particularly for our friendship. Without all of you, this work could not have been possible, hence accept by gratitude.

Introduction

We have a common, natural tendency to consider spiders, millipedes and woodlice as insects. None of them belong to this group: each of these animals belongs to distinct classes of the animal kingdom. The ensemble of these invertebrates belong, on the other hand, to the same phylum, that of the arthropods. This book does not only outline insects but also other arthropods.

BIODIVERSITY

Arthropods are by far the most abundant animals on our planet. To underscore their significance, suffice it to say that of the nearly 1.5 million species of plants and animals described up till now in our world, about 1.2 million are arthropods, among which there are 950,000 species of insects! It is difficult to provide exact figures, since with the passage of each year, new species are discovered. Therefore, more than 5,000 insect species are described each year. To estimate the number of species currently in existence is thus a very tricky exercise. The numbers given encompass a broad range from 6 million to 50 million.

Arthropods: What are they and how many are there?

We give the name arthropods (*arthro* = articulation and *pod* = foot) to all invertebrates with an external skeleton (exoskeleton) and with articulated appendages.

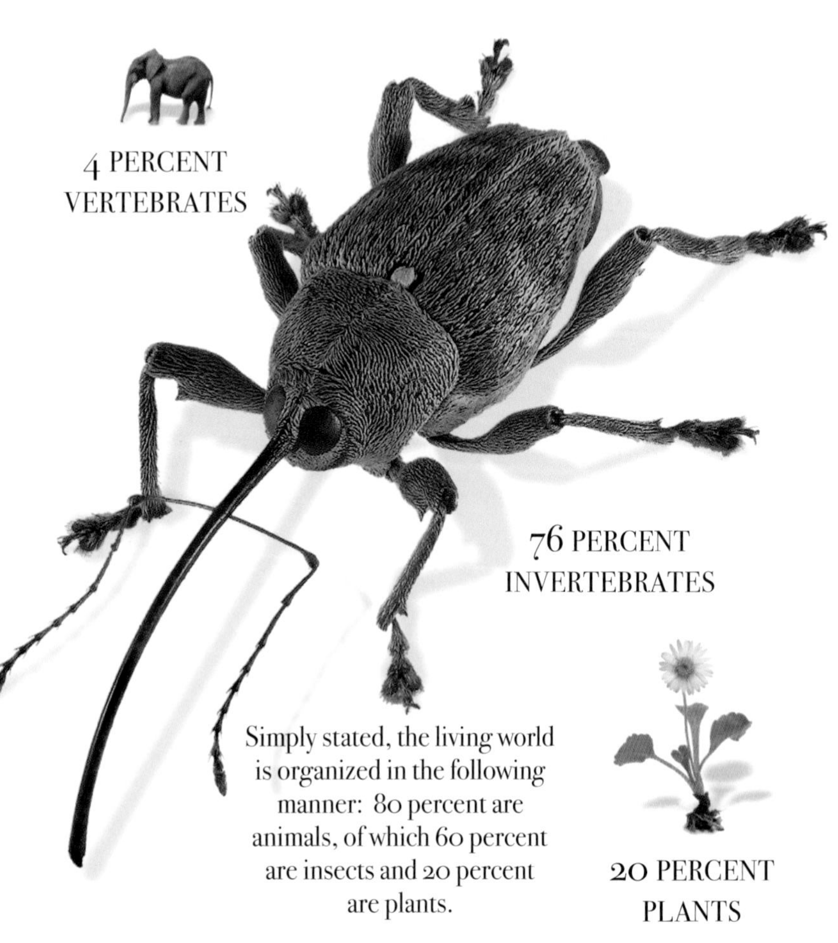

Simply stated, the living world is organized in the following manner: 80 percent are animals, of which 60 percent are insects and 20 percent are plants.

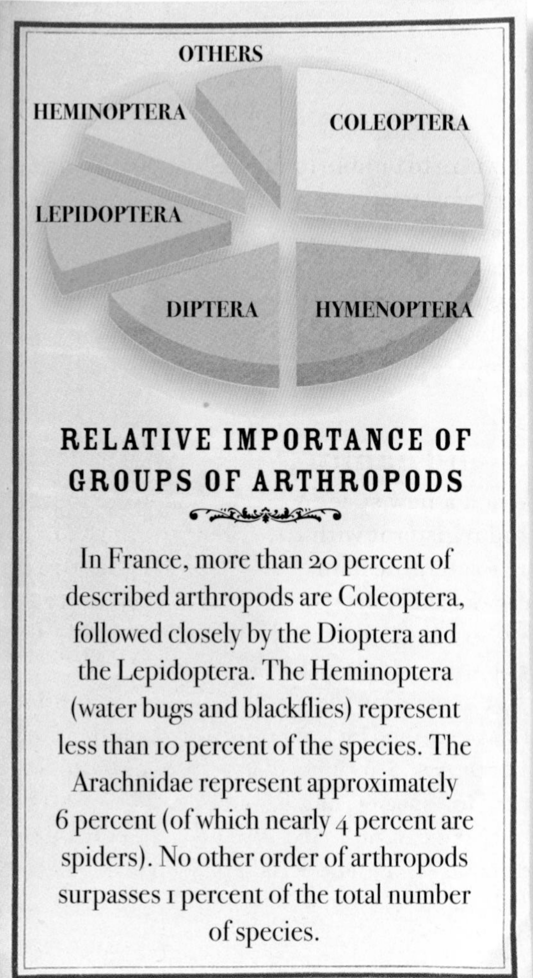

RELATIVE IMPORTANCE OF GROUPS OF ARTHROPODS

In France, more than 20 percent of described arthropods are Coleoptera, followed closely by the Dioptera and the Lepidoptera. The Heminoptera (water bugs and blackflies) represent less than 10 percent of the species. The Arachnidae represent approximately 6 percent (of which nearly 4 percent are spiders). No other order of arthropods surpasses 1 percent of the total number of species.

CLASSIFICATION OF LADYBUGS

Kindom: *Animalia*
Subkingdom: *Eumetazoa*
Phylum: *Arthropoda*
Superclass: *Hexapoda*
Class: *Insecta*
Order: *Coleoptera*
Suborder: *Polyphaga*
Infraorder: *Cucujiformia*
Superfamily: *Cucujoidae*
Family: *Coccinellidae*
Subfamily: *Coccinellinae*
Tribe: *Coccinellini*
Genus: *Coccinella*
Subgenus: *Coccinella*
Species: *septempunctata*
Subspecies: *septempunctata*

SYSTEMATICS

The birth of systematics

In order to organize the vast knowledge about species, it became necessary to give each a name to unambiguously identify and classify them. The Swedish botanist Carolus Linnaeus (1707–1778) thought, and rightly so, that ignoring the name of things contributed to having less knowledge about them. Thanks to his ordered way of thinking, he organized, mostly on the basis of anatomical criteria, the branches of natural history in his book *Systema Naturae*. He established the foundations of zoological nomenclature and, more significantly, introduced binomial nomenclature. Though plant species dominate his work, he also described some 3,000 insects. Johann Christian Fabricius (1745–1808) continued the work of the Father of Systematics, and went on to describe more than 10,000 insect species on his own.

Classifying ...

Systematic hierarchy is based entirely on taxonomical rankings. Therefore, for the Seven-Spotted Ladybug, the classification is established as illustrated in the drawing to the left.

The impressive number of classification categories may appear overdone at first and pointless to a many of us. However it remains critical to fully describe, in a precise fashion, a ladybug from the considerable number of other like species in the living world.

Note that the suffixes (–oidae, –idae, –inae, etc.) of the taxonomic names inform us of the ranking of that name. It is rare to be able to identify a species with complete certainty. For the most part, you must use a word that refers to the family, with an identifying key, to arrive with any degree of certainty to naming a specimen. Investigation is often undertaken with a microscope, scrupulously noting all identifying characteristics. The first time around, progress among the different species is slow and fastidious. For certain groups, identification is only possible by comparing the external genitalia of related species.

Carolus Linnaeus, 1707-1778

... and naming

When a new species is discovered, its discoverer will give it a scientific name and register it with the International Commission on Zoological Nomenclature. The scientific name, established along the principles of binomial nomenclature, is composed of the genus name (most often of Greek origin) followed by the name of the species (in Latin). This name is then followed by the name of the discoverer and the year of the description.

For example, let us consider the cetonian known as the Rose Chafer. This species was described by Linnaeus in 1761 under the name ***Scarabaeus auratus* Linnaeus, 1761**. But in 1775, Fabricius distinguished the cetonians from other scarabs and assigned a new name to the genus: Cetonia. The species' name remained the same, though it was conjugated to agree with the feminine name preceding it, to become *aurata*. To note such a modification, it was agreed to place the name of the discoverer and the date of his discovery in parentheses. Consequently, the scientific name of the Rose Chafer will from now on be ***Cetonia aurata* (Linnaeus, 1761)**.

During Linnaeus' time, Latin was the common scientific language. Today, we continue to use both Latin and Greek for scientific names. This nomenclature offers the quality of universality: the scientific name allows one to designate, without ambiguity, the same species in all countries of the world.

... and phylogeny

The traditional mode of classification, used since Linnaeus' time, was established along almost exclusively morphological criteria. By a fortuitously good choice we have since established a classification system that also provides a fairly accurate picture of the parental links between living species. In fact, the morphology of a species is determined by its genes. Therefore, a bee more closely resembles a wasp than it does a cow since the last common ancestor shared between bees and wasps is much less distant in the past than the last common ancestor (there surely was one!) of bees and cows.

However, occasionally these similarities might be misleading or even erroneous, since certain organisms may look alike without necessarily having a close common ancestor. The important progress made in the field of genetics (one can now "read" the DNA sequences of most living organisms) has provided the means to correct certain errors made when taking a solely morphological approach to classification. For example, it has been shown that crocodiles are in reality closer genetically to birds than to lizards though, at first glance, a crocodile would be thought to resemble an iguana much more than a robin! Modern classification, which is uniquely founded on parental links brought on through evolution (a classification system called phylogenetic) continues to evolve from day to day, in light of the latest discoveries. Therefore, contrary to general opinion, classification is very much a vital, living, dynamic science, which utilizes the most modern technological tools to study the living world. It is, however, a science that stills respects its origins, acknowledging that its foundation is based on data methodically gathered over three centuries.

Classification of the Book's Arthropods

We have just described how systematics are constantly being reworked and are evolving. The following described classifications are based on classical reference points, while being cognizant of recent phylogenetic hypotheses.

THE CHELICERATES

[CLASS ARACHNIDAE]

The arachnids are chelicerate arthropods. Contrary to insects, they have four pairs of legs. Their bodies are divided into two parts: the cephalothorax (head and thorax are fused as one) and the abdomen.

They have three types of articulated appendages:
- **The movable legs:** four pairs, used mainly in locomotion (and the manipulation of silk in spiders).
- **The pedipalps** (*pedi* = foot and *palpe* = to touch) are tactile appendages. Equipped with sensory hairs, they are located close to the mouth and employed in the capture of prey. In scorpions, these are powerful pincers.
- **The chelicerata** (*chel* = pinch and *cer* = horn) are horned pincers (transformed into venomous hooks in spiders) that play the same role as the mandibles of insects. In ticks, the chelicerata form a girdle ending in claws that are used to pierce skin.

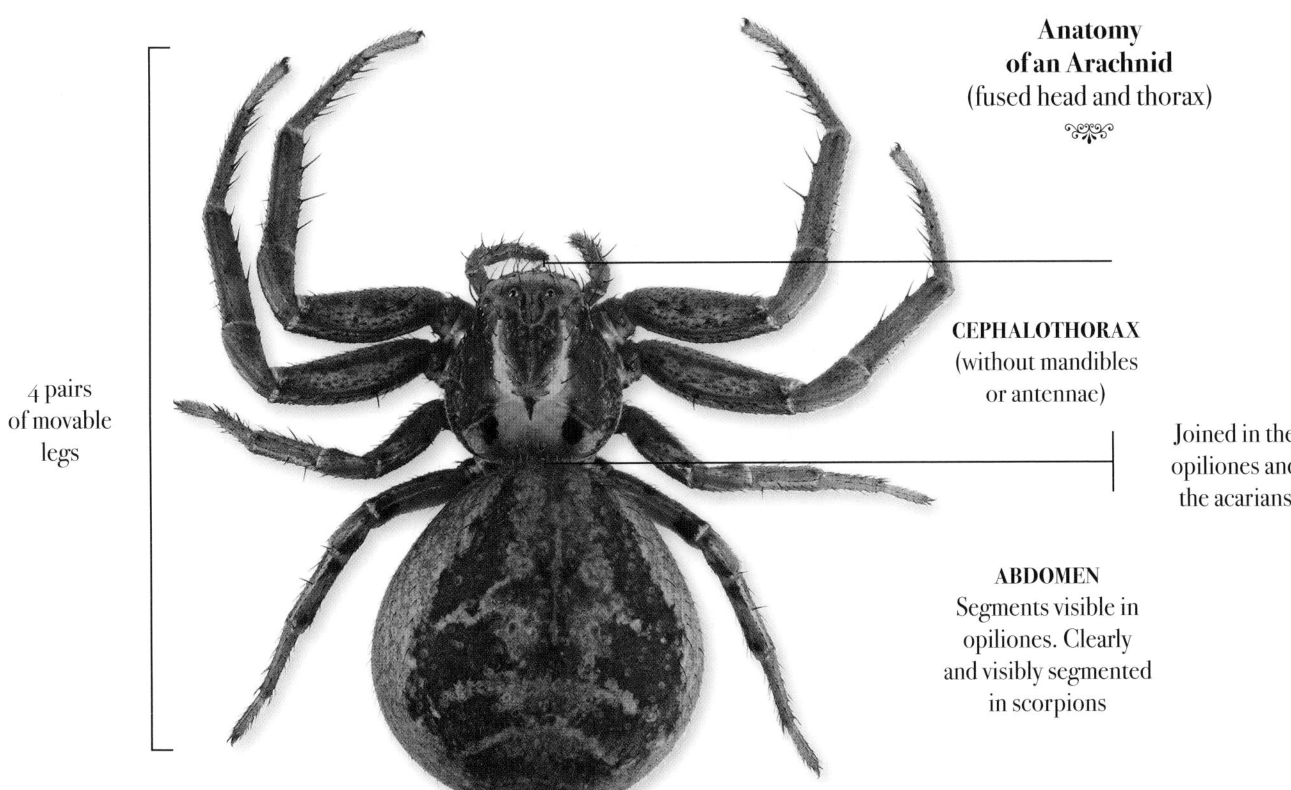

• Order Scorpionids

Scorpions are fairly large arthropods. They have one pair of noticeably visible pincers (the pedipalps) and small chelicerae around their mouth. The abdomen, which is often bent in position, ends with a fang and a stinger.

• Order Araneae

The Araneae are the spiders. Unlike insects, they have simple eyes (which are not multifaceted), most often four pairs in number, situated in two to three rows on the head. The number and location of the eyes allows us to distinguish between various families of spiders. The pedipalps of male spiders end in a bulbous copulator, which resembles boxing gloves. With a shape characteristic to the species, it inserts itself just like a key into the genital orifice ("the lock") of the female during reproduction.

• Order Opiliones

Often mistaken for spiders, the Opiliones, commonly known as harvestmen, belong to an independent order. Their small, globular bodies are composed of a joined cephalothorax and abdomen (this is not the case with spiders). Other dissimilarities with spiders include: the segmentation of the abdomen is clearly visible; their four pairs of legs are elongated and spindly; and they only have two large eyes set atop a prominent tubercle.

• Ordre Acarians

The Acarians are part of a numerically significant order (more than 1,200 species) among the arachnids. Their cephalothorax is fused to the abdomen and their body has a rounded shape. They are often quite small – almost invisible to the naked eye. Ticks are larger sized Acarians.

THE MYRIAPODA

The word Myriapoda (*myriad* = ten thousand and *pod* = foot) was created by Latreille in 1796 to designate the group of millipedes, though we know of no species that actually has a thousand legs. These animals are rarely seen since they shy away from the light. Their bodies are multisegmented, though we cannot distinguish a specific thorax or abdomen. Like insects, they have mandibles and not chelicerata. The Myriapods are divided into two classes: the **Diplopoda** and the **Chilopoda**.

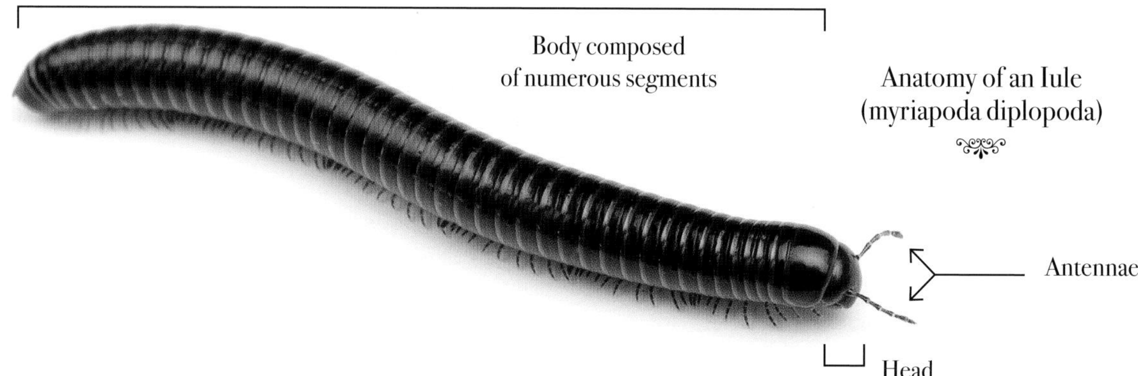

Anatomy of an Iule (myriapoda diplopoda)

[CLASS DIPLOPODA]

The Diplopoda (*diplo* = double and *pod* = foot) are myriapods in which each segment carries two pairs of legs. They are essentially vegetarians. Their genital orifices are situated generally at the front of the animal, behind the second pair of legs. They practice direct, internal fertilization. When moving, the right and left legs of the same segment travel simultaneously.

p. 128

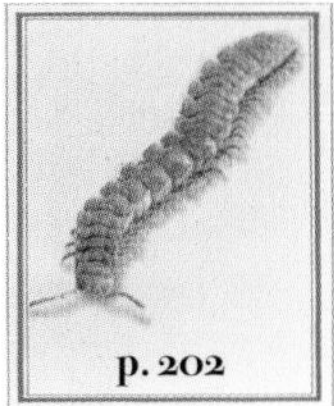
p. 202

[CLASS CHILOPODA]

The Chilopoda (*chilo* = border and *pod* = foot) only have one pair of legs on each segment. The first pair of legs, located near the mouth, are modified as powerful venomous fangs (prehensors) that these predators can use to pierce the skin of their prey. Their genital orifices are found at the back of the animal. These arthropods practice external or indirect fertilization with the help of a spermatophore. During locomotion, the right and left legs of the same segment move independently.

p. 160

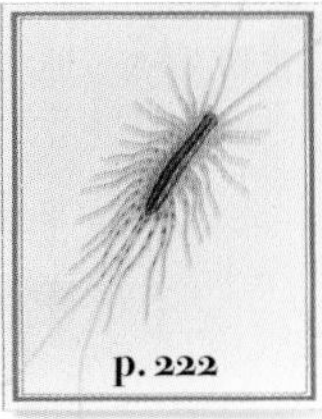
p. 222

THE CRUSTACEANS

[CLASS MALACOSTRATA]

The malacostratans (*malakos* = soft and *ostracon* = shell) constitute one of the five classes of crustaceans. By far the most numerous in terms of species, this class includes the well-known marine crustaceans (lobsters, crab, shrimp, and so on) as well as the water lice (amphipods) and the woodlice (terrestrial isopods). The woodlice have two pairs of antennae (of which two very small ones are called antennules). Surprisingly for terrestrial animals, they breathe through brachia (located at the level of their legs). This is why they need humid environments to survive.

Anatomy of a Woodlouse

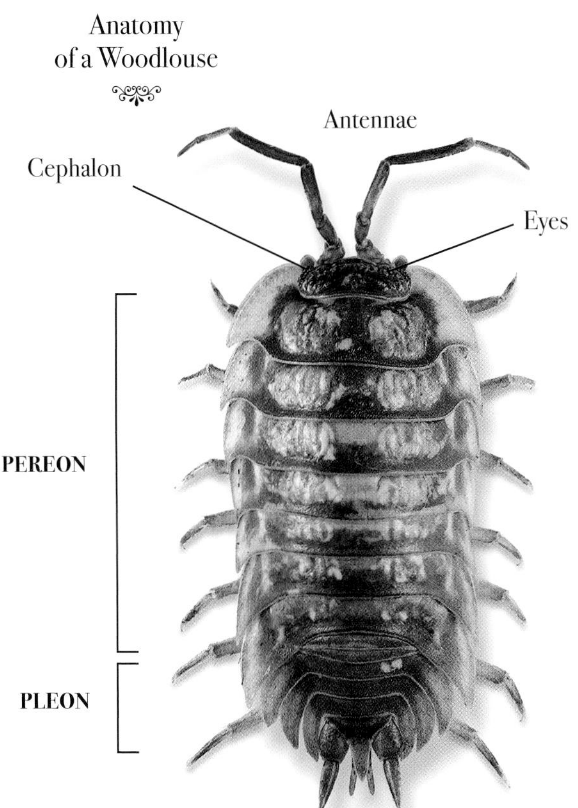

The body of a woodlouse is divided into three parts:

- **The head** (cephalon), which contains the eyes, mouthparts and two pairs of antennae.
- **The thorax** (pereon) is composed of seven segments, each equipped with one pair of legs (pereipods).
- **The abdomen** (pleon) is composed of six segments, which contain both the respiratory and copulatory appendages.
- **The two terminal appendages** (uropods) are equivalent to the flattened appendages that form the tail and serve as swimming devices in crayfish.

• Order Isopoda

The terrestrial isopods (*iso* = equal, and *pod* = foot) have **seven pairs of identical legs**, from which they derive their name, and represent the terrestrial crustaceans commonly known as woodlice.

These arthropods are closer in resemblance to shrimp than to insects!

p. 186

THE HEXAPODS

[CLASS INSECTA]

The insects (*in* = inside and *sect* = cut, equivalent to the Latin *entomo*) have bodies divided into three parts (the head, the thorax and the abdomen). They all have three pairs of legs, one pair of antennae and, for the most part, one or two pairs of wings. They are very numerous.

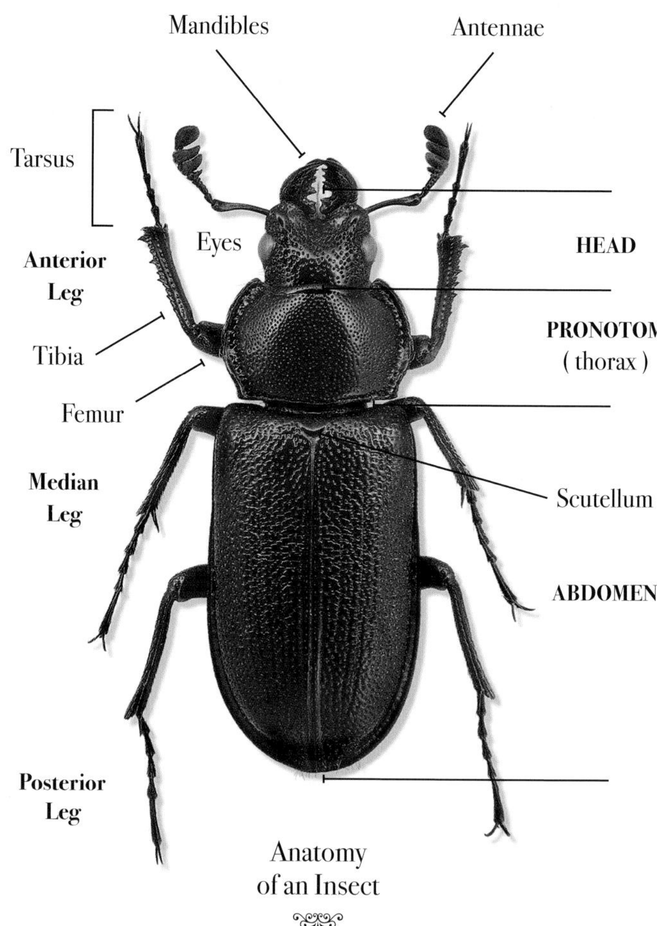

Anatomy of an Insect

• Order Zygentoma (also known as order Thysanura)

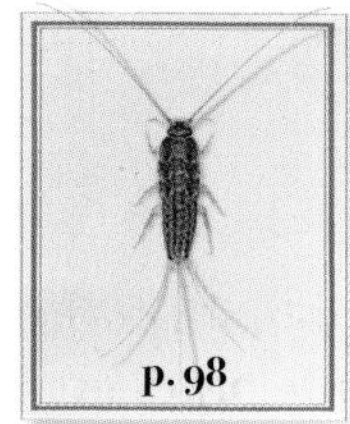
p. 98

The silverfish comprise, along with the firebrats, the ancient order of the Thysanura (*thysano* = fringe and *ur* = tail). These two groups are now considered as entirely separate orders – with the Archeognathes on one hand and the Zygentomes on the other . However, the term Thysanura is still used to designate the latter two. The development of the Zygentomes is ametabolic – that is, it occurs without metamorphosis, through successive molts. One remarkable trait is that the adults continue to molt regularly throughout their life cycle. These are primitive insects that never developed wings (apterygotes).

• Order Odonata

The odonates (*odo* = tooth and *nate*-, from *gnath* = jaw or mandible) together comprise the damselflies and dragonflies. These are often very large insects, with a long, thin body, capable of rapid flight. Their development is known as hemimetabolic: the larvae have a different living environment (they are aquatic) than their adult counterparts, and metamorphosis is incomplete. These are frightening predators both in their larval and adult stages. The larvae are equipped with an extendible oral apparatus that is also articulated (the "mask"), outfitted with pincers to capture their prey. The order Odonata is subdivided into two suborders: the **Zygoptera** and the **Anisoptera**.

SUBORDER ZYGOPTERA

The Zygoptera (*zygo* = pair and *pter* = wing), or damselflies, have their wings stuck together when at rest, and an abdomen thinner than that of the dragonflies. Their widely spaced eyes give their head a transverse elongated appearance. Their larvae, thin and elongated, breathe and move around with the help of three foliated gills (tracheobranchiae) located at the end of the abdomen.

p. 58

p. 90

SUBORDER ANISOPTERA

The Anisoptera (*aniso* = uneven and *pter* = wing), or dragonflies, have a pair of different wings (the forewings are smaller), spread out on each side of the body at rest. The large larvae, flattened and squat, breathe via tracheobranchiae hidden in the posterior intestine. They propel themselves by shooting water outside via violent contractions of their muscular intestinal wall.

p. 156

• Order Blattodea

p. 50

Until recently, the Blattodea were included with the cockroaches and the mantises (below) in the order Dictyoptera. These insects share the common presence of a complex, asymmetrical copulation organ and the fact that they lay their eggs in an ootheca. Their larvae have the same life cycle as those of adults (the development is of the heterometabolic type). We are most familiar with the cockroaches that choose to share our homes, though many among them live in the wild.

• Order Mantodea

p. 170

The mantids have an elongated shape. They have forelegs adapted for capturing their prey. These seizing legs have tibias in the shape of slats, which fold onto themselves, much like a pocketknife, on their femurs. Mantids love warmth. As with cockroaches, young mantids resemble the adults (heterometabolic development).

• Order Orthoptera

The orthopterans (*ortho* = right and *pter* = wing) are equipped with posterior legs adapted for jumping and mouthparts capable of grinding food. The Orthoptera display the particular trait of having chirring sound-making organs to emit sounds, and auditory tympana specialized to capture these "chants." If not for their smaller size and absence of wings, juveniles are nearly identical to their adult counterparts (heterometabolic development). The order Orthoptera is divided into two suborders: **Ensifera** (grasshoppers) and **Caelifera** (crickets).

SUBORDER ENSIFERA

The ensiferans (*ensi* = sword and *fer* = to carry) have long antennae and all are omnivorous. The female has a well developed egg-laying tube (ovipositor) shaped like a sword. In several species, mating is carried out via an intermediary vesicle (spermatophore) that contains sperm. The auditory organ (tympanum) is located on the tibia of the forelegs. The chirring sound-making organ is located on the wings.

p. 228

p. 136

p. 134

SUBORDER CAELIFERA

The caeliferans (*cael* = cavity and *fer* = to carry), unlike the ensiferans, display an auditory cavity on the side of their abdomens. Their antennae and ovipositors are short. The chirring sound-making is accomplished by rubbing the femurs of the posterior legs on the raised band of the wing case. Their diet is almost exclusively herbivorous (they consume mostly grains).

p. 184

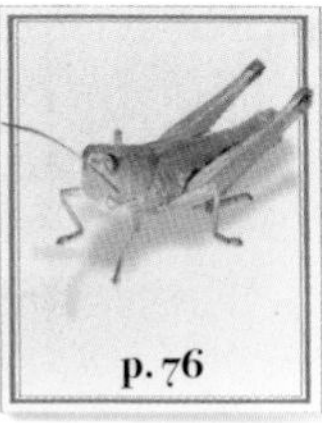
p. 76

• Order Dermaptera

p. 122

The forficulae (or earwigs) form the small order of dermapterans (*derm* = skin and *pter* = wing). They are recognizable by small circles in the shape of pincers at the extreme end of their abdomens, and very short wing cases. They are photophobic (have an aversion to light), and are equipped with grinding mouthparts to feed on plants and small animals. Earwigs carry out parenting duties with their young, a trait very uncommon in insects that don't live in colonies. Their metamorphoses are incomplete (heterometabolic).

• Order Hemiptera

The hemipterans (*hyme* = half and *pter* = wing) originally represented the order of bugs. They are all equipped with a beak that is is capable of both pinching and sucking. There are many suborders. Hemipteran metamorphosis is incomplete (heterometabolic).

SUBORDER HETEROPTERA

Families: Gerridae, Pentatomidae, Coreidae, Pyrrhocoridae and Tingidae

The Heteroptera (*hetero* = different and *pter* = wing) resemble bugs. Their forewings, or hemelytra, are comprised of a hard basal part and a membranous extremity, vestiges of the original membrane. The scutellum is very well developed.

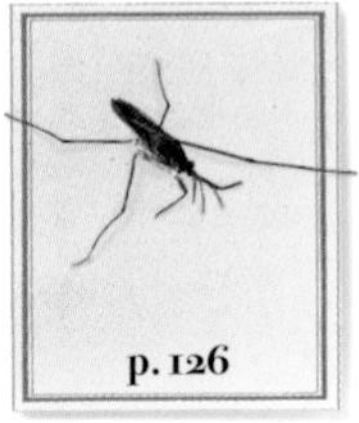
p. 126

p. 192

p. 118

p. 132

p. 92

p. 208

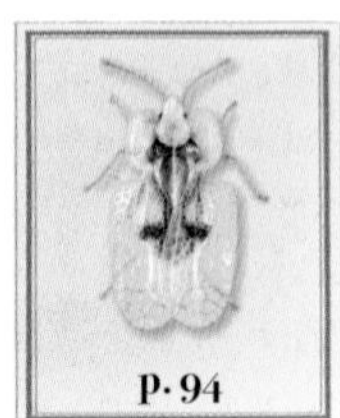
p. 94

SUBORDERS AUCHENORRHYNCHA AND STENORRHYNCHA

Families: Cicadidae, Membracidae, Cercopidae and Aphididae

These represent the ancient suborder Homoptera (*homo* = similar and *pter*-wing), which is now separated into several suborders. Auchenorrhyncha regroups the cicadas, hoppers and treehoppers. Order Stenorrhyncha includes aphids and whiteflies. Their wings are homogenous in structure and folded in a tentlike fashion. All feed by sucking their nutrients from plant sap.

p. 84

p. 68

p. 72

p. 24

• Order Coleoptera

The order of coleopterans (*coleo* = case and *pter* = wing) is the most diverse group in the animal kingdom. They derive their name from the fact that the anterior elytra (hardened forewings) are perfectly joined, forming a sort of case that covers the membranous hindwings. Many coleopterans have lost the ability to fly. All have grinding mouthparts. Coleoptera develop through complete metamorphosis (holometabolic development); they go through an immobile stage, called the nymph, during which the larvae transform into adults.

Families: Carabidae, Cicindelidae, Dytiscidae and Gyrinidae

The tiger beetles distinguish themselves from the ground beetles by their large, prominent eyes. Their larvae hunt, hidden in a well that they dig themselves. The Carabidae are ferocious carnivores. The majority of ground beetles are nocturnal and incapable of flying. The Brachininae, which belong to the family Carabidae, are numerous and include the famous bombardier beetles. The predaceous diving beetles are large, aquatic and fearsome carnivores. Their bodies, perfectly designed for an aquatic existence, has hindlegs adapted for swimming. The gyridins, for their part, are very small coleopterans that quickly slide along the surface of the water and have the particular trait of divided eyes – to see both above and under the water.

p. 86

p. 62

p. 64

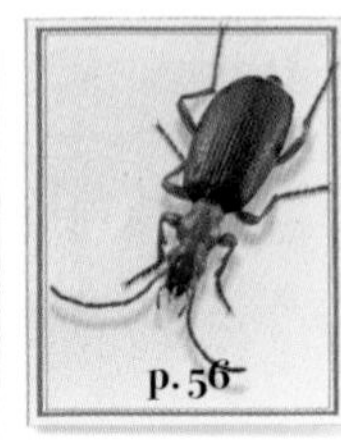
p. 56

p. 110

p. 138

Families: Staphylinidae and Silphidae

The Staphylinidae have an elongated body with small elytra that exposes part of their abdomens. They represent, after the weevils, the richest grouping of coleopterans. The carrion beetles, which belong to the family of Silphidae, contribute to the decomposition of cadavers. The care they bring to feeding their offspring is truly remarkable.

p. 112

p. 180

Families and subfamilies: Lucanidae, Geotrupidae, Melolonthinae, Dynastinae and Cetoniinae

Among the Lucanidae, the stag beetles are notable for their large size and, in particular, the formidable mandibles found in the males. The dynastins, also quite large, are represented in France by, among others, the rhinoceros beetles. The Geotrupidae, which have a very rounded shape, have large burrowing legs, which they use in their contribution to recycling dung. The June bug, one of the most recognizable insects, belongs to the subfamily Melolonthinae. The Cetoniinae, the Prune-stinger and the Trichies are Cetoniinae.

p. 162

p. 32

p. 238

p. 174

p. 144

p. 188

p. 74

p. 190

p. 236

Families: Buprestidae, Elateridae, Cantharidae and Lampyridae

The buprestids (jewel beetles) and elaterids (click beetles) have similar body shapes. The former display superb metallic colors; the latter distinguish themselves by the ability to jump when placed on their backs.Their pronotum (thorax) is characterized by prominent posterior angles. The Cantharidae are usually associated with umbellifers, a family of aromatic plants. The well-known fireflies, famous for being luminescent, belong to the family Lampyridae.

p. 34

p. 28

p. 210

p. 152

Family Chrysomelidae

All chrysomelids are plant eaters, and several, such as the Criocerinae, potato beetles and flea beetles, are known and feared by gardeners. Several species are brightly colored.

p. 158

p. 232

p. 154

p. 80

p. 204

p. 142

p. 66

Family Cerambycidae

The cerambycids, better known as longhorn beetles, are insects that may be quite large. As the name indicates, the noticeable attribute of the longhorn is the size of its antennae (much more noticeable in males). Their larvae bore tunnels in trees. Two species illustrated in this book are nationally protected.

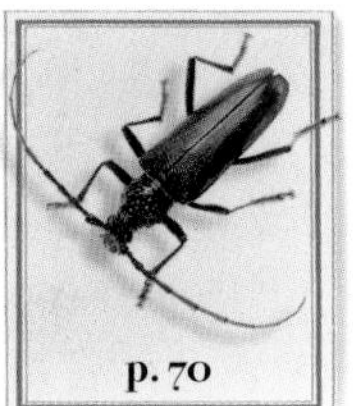
p. 70

p. 216

p. 218

Families: Coccinellidae, Meloidae, Oedemeridae, Pyrrhocoridae and Tenebrionidae

Contrary to popular belief, the spots on the coccinellids (ladybugs) are not an indication of the insect's age. The meloids are large insects with a narrow pronotum. Their larvae develop as hymenopteran or orthopteran parasites. Members of the small family of oedemerids sometimes have deformed elytra and, in males, the hind femurs are hypertrophied. The pyrrhocorids, very few in number, live in woodlands. Their predatory larvae lie under the bark of trees. The vast majority of tenebrionids value warmth and dryness; some live in desert environments. Some species of the Tenebrionidae are pests, consuming flour and cereals.

p. 88

p. 172

p. 182

p. 206

p. 226

Families: Rhynchophoridae, Curculonidae, Dryophthoridae and Scolytidae

p. 102

p. 100

p. 212

p. 148

The very numerous weevils, of diverse morphology and size, are nevertheless recognizable by their elongated heads and a rostrum, at the end of which are located their small mandibles. The rhynchophorids comprise those insects known as cicadas, and the curculonids, the snout beetles. The Dryophthoridae species highlighted in this book is the Red Palm Weevil. The scolytids have an elongated, cylindrical body, perfectly adapted to scurry in the tunnels they bore into trees.

• Order Neuroptera

Family Chrysopidae and superfamily Myrmeleontoidae

p. 82

p. 116

Larvae of the neuropterans (*neuro* = nerve and *pter* = wing) are all predatory. In the doodlebugs and lacewings, the oral opening is closed. The larvae inject their saliva and aspirate the liquefied tissues of their prey with their deep, hollow mandibles.

The doodlebugs (also known as antlions) could be mistaken for dragonflies, although an examination of their swollen antennae reveal the difference between these groups of insects. The lacewings are particularly interesting for the voracious appetite of their larvae. They are used commercially to combat aphids. The larva does not resemble the adult and its metamorphosis is complete (holometabolic development). This order also includes the snakeflies, alderflies and owlflies.

• Order Hymenoptera

The order Hymenoptera (*hymen* = membrane and *pter* = wing) is the second most species-rich order after Coleoptera. There are 8,000 species in France and it is estimated that several remain to be discovered (which is not the case for the coleopterans). Hymenopterans have two pairs of wings. Half of the species are found in the Braconidae and Ichneumonidae families. The development of hymenopterans (for which certain life cycles are very complex) is as a complete metamorphosis (homometabolic). The order Hymenoptera is divided into two suborders: **Symphyta** and **Apocrita**.

SUBORDERS SYMPHYTA AND APOCRITA

The body of symphytes (sawflies) has an unusual feature: all of the insect's reproductive organs thread through a thin petiole up to the abdomen. The apocrites are very numerous and are subdivided, in turn, into two groups: the **terebrants** and the **aculeates**.

• THE TEREBRANTS

Families: Cynipidae and Ichneumonidae

The terebrants are parasitic hymenopterans. The females have an egg-laying tube (ovipositor), which is generally thin and long, designed for the specific purpose of inserting their eggs into a host, which can be either a plant or an animal. Their ability to parasitize other arthropods gives them a major role to play in the latter's population control.

p. 30

p. 104

p. 48

p. 214

• THE ACULEATES

Families: Chrysididae, Apidae, Vespidae and Formicidae

These are the most well known hymenopterans. Many live in colonies. The females generally lack an ovipositor, though they do have a stinger linked to venom glands.

p. 78

p. 246

p. 52

p. 36

p. 242

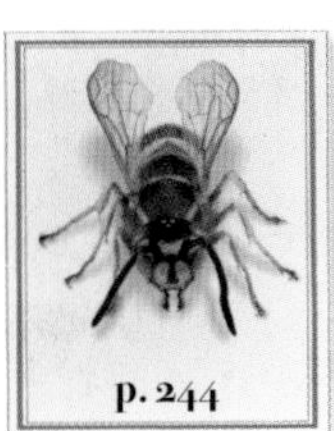
p. 244

p. 124

• Order Mecoptera

Family Panorpidae

The mecopterans (*meco* = long and *pter* = wing) comprise a very small order which includes the Panorpidae, or scorpionflies. There are only nine species in France. Mecopterans resemble neuropterans, though they can be distinguished by a long oral apparatus in the form of a muzzle which ends in masticating jaws. Their metamorphosis is complete (holometabolic). The male's genitals look like a scorpion's tail.

p. 194

• Order Diptera

Families: Tipulidae, Tabanidae, Drosophilidae, Calliphoridae, Muscidae, Bombylidae and Syrphidae

The order Diptera (*di* = two and *pter* = wing) consists of an important number of diverse insects. The dipterans are characterized by one pair of wings, the vestiges of a second pair remaining as visible polelike structures called halteres. Since the order contains both flies and mosquitoes, it has a very bad reputation. However, the Diptera are very useful to humankind and to the biosphere: the saprophagous and phytophagous species work to aid in the elimination of organic substances; and the predatory and parasitic species contribute to the regulation of other species. Development is holometabolic, with a complete metamorphosis (worm/maggot, pupae and adult).

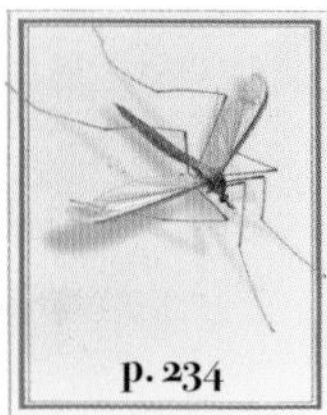
p. 234

p. 140

p. 108

p. 164

p. 178

p. 54

p. 114

• Order Lepidoptera

The order Lepidoptera (*lepido* = shell and *pter* = wing) gathers together the most appreciated and popular of all insects: the butterflies and moths. Their metamorphosis, also well known, is holometabolic (egg, caterpillar, chrysalis and adult). They have unique sucking mouthparts, called a proboscis, which allows them to draw nectar from flowers. A large portion of the species (3,000) are microlepidopterans. It is estimated that several more species are yet to be discovered. The lepidopterans are divided into two groups: the **rhopalocerans** and the **heterocerans**.

THE RHOPALOCERANS

Families: Papilionidae, Pieridae, Lycaenidae and Nymphalidae

The rhopalocerans (*rhopalo* = massive and *cer* = horn) encompass the large majority of daylight butterflies and moths. They are generally vibrantly colored, and have fine antennae ending in clumps.

p. 196

p. 200

p. 130

p. 168

p. 146

p. 240

p. 26

p. 44

THE HETEROCERANS

Families: Zygaenidae, Thaumatopoeidae, Sphingidae, Arctiidae and Gracillariidae

The heterocerans (*hetero* = different and *cer* = horn, meaning antennae) have variably shaped antennae. They encompass the nocturnal butterflies and moths. Despite this, the smoky moths (Zyaenidae) and the sphynx moths (Sphingidae) are diurnal species.

p. 248

p. 230

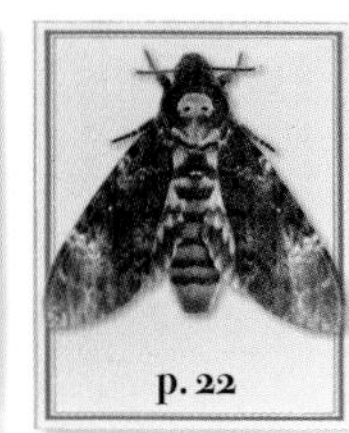
p. 22

p. 166

p. 40

p. 60

Development and sensory organs in arthropods

SKELETON AND DEVELOPMENT

While vertebrates have an external skeleton surrounded by muscles that attach to it, arthropods are quite different. They do not have a spine, and their legs and other appendages are comprised of hollow articulated segments, activated internally from the muscles. To observe an insect is indeed to observe its skeleton! This external skeleton (exoskeleton) is called the cuticle. Composed of chitin (an amino sugar), the cuticle is more or less rigid according to each species. A veritable armor that protects the animal, it nevertheless offers a major inconvenience – that of not being flexible. To grow, an arthropod must therefore molt, that is, cast off its old "armor" to build a larger new one, with all the risks that this process entails.

In this way, the development of arthropods occurs in a discontinuous manner: the animal goes through a succession of stages during which their size is fixed and the cuticle rigid, interspersed with molts during which the cuticle is soft and briefly allows a growth spurt. Molting is triggered by hormones. The animal ceases to feed itself and becomes motionless. The old carapace splits and peels away to reveal a new one, still quite soft and not pigmented. To save on "raw materials," a large portion of the old cuticle is used to manufacture the new one; from the old one, there remain only cast skin called exuviae. The new cuticle must first harden before the arthropod can fully regain all its functions. This is why arthropods choose a site both discreet and shielded from predators to accomplish their molts.

All insects molt . . . but not all have the same type of development

For some, the juveniles resemble their adults, while others go through significant transformations, or metamorphoses (*meta* = following or, by extension, change; and *morph* = form) during their development.

• **THE AMETABOLITES**

The ametabolites (*a* = without and *metabol* = change/metamorphosis) do not go through metamorphosis. The juvenile is an identical, miniature version of the adult. This development occurs in silverfish (Lepistimatidae) and springtails (Collembulae). The adults continue to molt throughout their lives.

• **THE HETEROMETABOLITES**

These are also referred to as having "an incomplete metamorphosis." After the juveniles hatch from eggs, heterometabolitic (*hetero* = different and *metabol* = change/metamorphosis) insects share the appearance and, in general, the lifestyle of adults, though they are smaller and lack wings. Molting ceases in adulthood. The cockroaches and mantises, crickets and grasshoppers, earwigs and bugs (*see following illustration*) are all heterometabolic.

Dragonflies are heterometabolites that characteristically develop in two separate environments: the larvae are aquatic, whereas adults breathe air. These insects are said to be hemimetabolites (*hemi* = half and *metabol* = change/metamorphosis).

The Development of Heterometabolites

The Development of Holometabolites

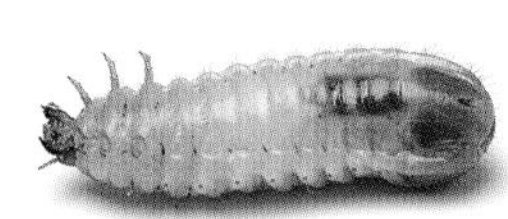

• THE HOLOMETABOLITES

These are also referred to as having "complete metamorphosis." The vast majority of insects are holometabolic (*holo* = complete and *metabol* = change/metamorphosis). A profound transformation is seen from the juveniles to adults. This transformation is not only anatomical, but also, for the most part, affects the insects' lifestyles. The larva grows and molts with each stage, and also goes through an intermediary, immobile phase before reaching the adult stage. The latter is called the nymph phase. When it reaches its full-grown size, the larva (referred to as a caterpillar in butterflies, a maggot in flies, and a worm in beetles) becomes immobilized (often in a previously constructed shelter), after which it molts into a non-motile nymph. The shelter may be in the form of a clump of soil, a silk cocoon or a pupa in the dipterans, or a chrysalis in the lepidopterans. During its final molt, the nymph becomes an adult – also called an "imago" (see following illustration). This type of development is common to all coleopterans, lacewings and doodlebugs, hymenopterans, scorpionflies, dipterans and butterflies.

HOW ARTHROPODS EXPERIENCE THEIR WORLD

• SIGHT

Very much jewel-like, the eye in many arthropods is a surface composed of a multitude of facets. In fact, it's composed of a veritable bouquet of simple eyes (the ommatidium), similar to tubes seen in cross-section, packed tightly against one another like bunched asparagus. Each "facet" on the eye's surface is, in reality, a lens of one of the ommatidia. These ommatidia vary in number according to the species. Woodlice only have a few dozen, flies have several thousand, whereas some dragonflies (Aeshenidae) have tens of thousands. The perceived image as viewed by an arthropod is therefore a composite one: it consists of as many "pixels" as it has ommatidia. Their color vision is also quite different from ours, and several arthropods can perceive ultraviolet light.

As well as these compound eyes, several species of arthropods have more simplified eye structures, referred to as ocelli. Positioned between the compound eyes, often in groups of three, the ocelli appear to be useful in perceiving light intensity as well as movements. Spiders only have these simple eye structures, very sensitive to movements.

• TOUCH AND OTHER SENSES

Cramped inside their cuticular carapace, the arthropods are nevertheless equipped with several sensory devices capable of perceiving a large amount of tactile information. For example, a cockroach has, at the end of its abodomen, a pair of cerci that are extremely sensitive to any air movement. The glomeris are special organs, situated on each side of the antennae, which provide feedback on the ambient humidity. Several tactile hairs are spread over the entire carapace of arthropods, and are particularly abundant on the legs.

• SMELL AND TASTE

If arthropods are not endowed with a true nose, most have more than enough antennae to make up for the loss (not all however – note in particular their absence in spiders). Antennae are highly endowed with olfactory receptors. While antennae are known to play an important olfactory role in some species, they also have a tactile function.

Taste can also be perceived via the antennae, or the oral palps, though some butterflies use their feet to sample the flowers before dipping their proboscis in them, and that is how flies taste the spilled sugar on the dinner table.

• HEARING

Having no tympanic membrane, most arthropods are deaf, though they have the capacity to perceive air currents. However, in species where males attract females by transmitting sounds (e.g., cicadas, crickets), hearing is good, and therefore a tympanic membrane exists even if it is often located in unexpected areas. Thus, in orthopterans in the suborder Ensifera, the tympanic membranes are located just under ... the knees! Finally, certain nocturnal butterflies can hear ultrasound very well, thereby avoiding capture by bats.

Overview of "our little animals"

Death's-head Hawkmoth

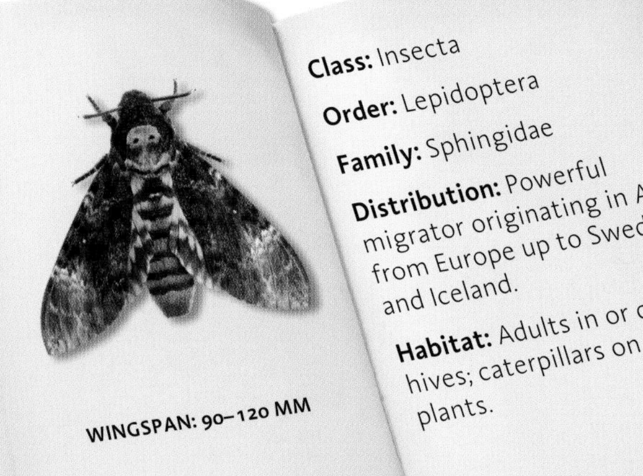

Origin of their name

The hawkmoths (sphinxes) owe their name to the body position, with their raised heads, often adopted by the caterpillars of the species. Some observers have drawn a parallel between this superior pose to that of the Sphinx of Thebes – and therefore the name. However, it's necessary to keep in mind that these first descriptions were often by keen proponents of ancient languages and literatures. *Archeontia* is the butterfly of the *Archeron*, which was, in Greek mythology, one of the rivers that the dead would cross to reach Hell. Equally morbid, *Atropos* is the name of one of the three Fates in charge of men's destinies and more particularly the Fate that cut the thread of life... With back markings resembling the Seal of Hell, the Death's-head Hawkmoth carries with it an omen foretelling death.

An otherwordly creature

Along with the Big Night Peacock, the Death's-head Hawkmoth is the largest European moth. It is rarely seen during the day in its adult form in our regions, since its behavior is essentially crepuscular (they love to gather pollen from jasmine flowers) and occurs only for a very short time span. The giant caterpillars, brilliantly colored or dull brown and the enormous chrysalis are much more often the subjects of observations by gardeners who hoe into their potato-shaped legs. Reaching 10 to 13 cm at maturity, the caterpillars develop on several underground plants (potatoes, belladonna, tobacco and henbane). The small horn, or *scolus*, which sits upright at the rear of the abdomen, is characteristic of Death's-head Hawkmoths.

Brown Form of the Caterpillar

The caterpillars do not survive the winter. However, this butterfly is a big traveler: all springtime adults live in the south of Europe and, more often, in Africa. They have winged their way across the Mediterranean and can at times migrate all the way up to Sweden or Iceland.

Bold pillagers

To carry out its long voyage, this robust moth requires one of the most energizing fuels that nature has to offer: a bee's honey. To this end, it has a short, stocky snout modified as a piercing organ (that of other sphinxes is normally quite long and fine to enable sucking the nectar out of the heart of the flower's corollas). Irresistibly drawn to the smell of the honey, they penetrate the beehives and puncture the alveoli. The bees do not take this lying down, however, and it is not rare to see them kill the moth and embalm it in a layer of propolis.

Hawkmoth chrysalis

AND IT CAN ALSO SQUEAL!

The Death's-head Hawkmoth appears to be the only living insect capable of emitting a cry from its "throat." This plaintive cry, which is similar to that of a mouse's squealing, is produced when the animal is excited or stressed (for example, when it is grabbed). Audible at a distance of about 130 feet (40 m), it probably fed the rumors concerning the supernatural nature of the sphinxes.

Young Caterpillar

Acherontia atropos

(Linnaeus, 1758)

Wingspan:
90–120 mm

Pea Aphid

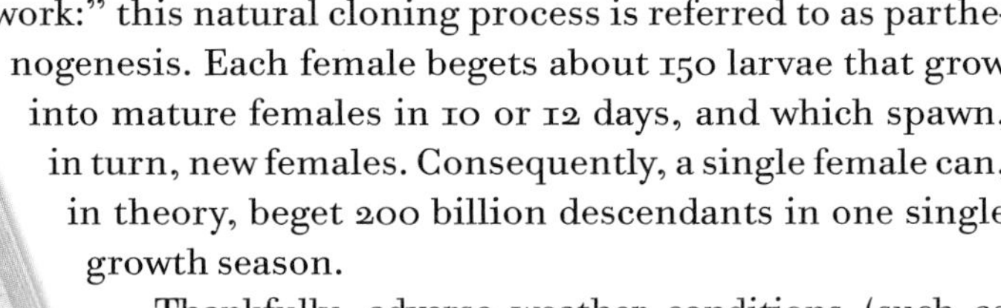

Aphid Larva

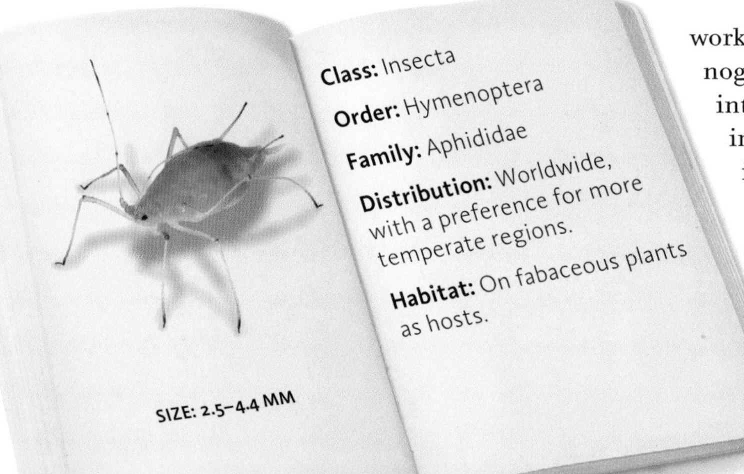

ORIGIN OF THE NAME

The word aphid ("small flea") immediately conjures up the image of feeding by suction, which these insects share with all Hemiptera. *Acyrthosiphon* describes the shape of the biting mouthparts: a deep, unbent tube/ siphon (*a* = without), and *pisum* describes their principal host plants: garden peas.

RELENTLESS SAP SUCKERS

With more than 4,000 species of aphids worldwide, some of which are almost cosmopolitan, only a few plants do not harbor at least one species of these profiteers. Once established on a plant, aphids do not move much – their sole activity is to relentlessly draw up sap in order to live ...

THE ATTACK OF THE CLONES

Aphids have an incredibly efficient reproductive strategy. They alternate an asexual reproduction, without fertilization, in spring and summer, with a sexual reproduction in fall (accomplished by the laying of eggs, thereby guaranteeing a winter survival). During the asexual phase, the "founding" females beget females genetically identical to themselves by birthing larvae "ready to work:" this natural cloning process is referred to as parthenogenesis. Each female begets about 150 larvae that grow into mature females in 10 or 12 days, and which spawn, in turn, new females. Consequently, a single female can, in theory, beget 200 billion descendants in one single growth season.

Thankfully, adverse weather conditions (such as droughts) and predators (Syrphae larvae or those of Chrysopae and Ladybugs) take care of controlling their spread. These precious garden friends, therefore, help limit the spread of aphids through biological warfare!

WITH OR WITHOUT WINGS

The pea aphid goes through its entire life cycle on fabaceous plants (such as pea, clover and alfalfa). Eggs hatch at the beginning of February and produce the "founding" females. Each one of them will produce several generations of wingless females through parthenogenesis. By mid- May, in response to the local over-population and the maturation of the host plant, winged forms begin to appear, free to go about colonizing other species of legumes. Such new colonies are composed once again of clones of wingless females. When days become shorter, a second winged generation makes its appearance and returns to the original host plant. By September, these females give birth to sexual individuals, males and females, who then couple. Each fertilized female lays some eggs, which will survive the winter.

THE BIRTHING OF "APHIDS"

In the midst of the Century of Enlightenment two authors became fascinated with the behavior of aphids. René-Antoine Ferchault de Réaumur (1683–1757) was one of the first to demonstrate that this insect was viviparous, that is to say capable of birthing completely formed offspring. Indeed he delightfully describes the "delivery" process of the "mother aphid." Not having observed either males or the act of coupling, he encouraged his close associate Charles Bonnet (1720–1793), a faithful follower, to conduct experiments to understand the process. Following many patient and minuscule experiments on aphids, "in isolation" in glass jars, he managed to demonstrate the existence of asexual reproduction in females which he referred to as "aphidettes!"

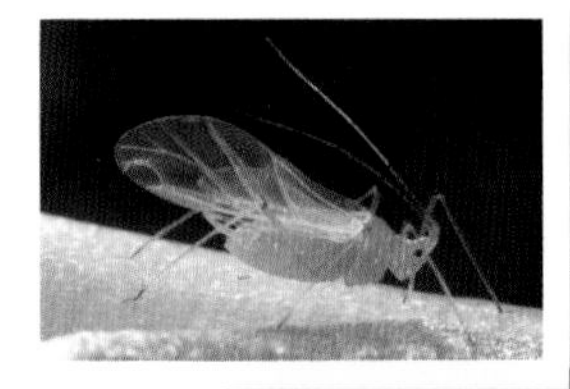

Winged female

A female "giving birth" ... to a larva

Acyrthosiphon pisum

(Harris, 1776)

Size: 2.5–4.4 mm

Aglais urticae

Small Tortoiseshell

An excellent camouflage: the underside of the wings

Class: Insecta

Order: Lepidoptera

Family: Nymphalidae

Distribution: Temperate zone of Eurasia.

Habitat: On embankments, at the edges of fields and gardens and on nettles.

WINGSPAN: 40–55 MM

Origin of the name

In Greek mythology, Aglaia (from the French Aglaé), was the youngest and most beautiful of the three Charities (the Roman Graces) – all daughters of Zeus. Personifying breathtaking beauty, Aglaé has inspired the name of this beautiful butterfly. The species name alludes to the host plant of the caterpillar, the nettle (*Urtica*). This species is sometimes confused with that of the Big Turtle (*Nymphalis polychloros*), a much larger but much less vividly colored butterfly, and frequents woodlands, borders and forest edges, where it can find its host plants (elms, willows, poplars, mountain ashes and so on).

Country living

The Small Tortoiseshell is a cousin of the Red Admiral Butterfly (p. 240) and the Peacock Butterfly (p. 146). It essentially develops on the stinging nettle (*Urtica dioica*), which grows in nitrate-rich soils, most notably on farmlands and other countryide areas. It is surprising to mostly encounter this species in rural areas. However, since nettle grows near all pastures and their enclosures, it is also possible to find this butterfly in high mountain pastures at an altitude of nearly 10,000 feet (3,000 m). (It is not very choosy and often flies in windy and cloudy weather conditions, while *Autochthon* species remain glued to the ground). Changes in farming practices, mowed-down embankments, the spread of pesticides and, more generally, the destruction of its host plant, have contributed to this insect's local decline.

Family life of small caterpillar

The female lays her large clutch of eggs on the underside of nettle leaves. The small warty caterpillars (see photo below) weave a collective tent under which they group-feed, and are protected from predators. They then scatter, grow and transform themselves into pretty golden chrysalises (following image) on the host plant. The adults burst forth from the beginning of June through August. They then gather wildflower pollen, as well as those of garden flowers – notably asters – and those of butterfly bushes (particularly the blue ones). Hibernation occurs in the adult phase in barns and other farm buildings, where many of them are devoured by spiders. By March or April, the survivors leave their dens to gather pollen from the willow catkins and dandelions.

Seduction on top, discretion below

The colors of the Small Tortoiseshell are a beautiful compromise between two dominant and contradictory necessities: to be visible to one's partner while not being seen by predators! Indeed, the upper surface of the wings, both in males and females, displays a pretty range of vibrant colors: orange-red, golden yellow, metallic blue, with a varying degree of black spots, and some white at the extremity of the forewings. The lower surface of the wings, on the other hand, is much more subdued, with an ornamentation designed to camouflage the butterfly when its wings are folded.

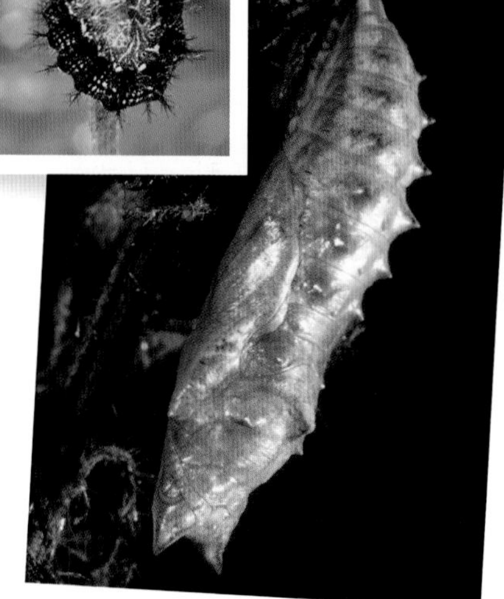

The caterpillar

No other chrysalis bears its name better (chrysos-, "gold") than that of the Small Tortoiseshell.

Aglais urticae

(Linnaeus, 1758)

Wingspan: 40-55 mm

Harvest Black-lined Click Beetle

Larva of the Harvest Click Beetle

ORIGIN OF THE NAME

In days gone by, we used to call "click beetles" miners, charged with undermining the walls of besieged towns. The insect was labeled with other common names, including blacksmith, tap-tap and click beetles. All these terms allude to the characteristic noise –a loud "click" – that these beetle make when righting themselves from lying on their backs. This startling behavior also explains the family name Elateridae, which has the same origin as the word "elastic." *Agriotes* means "rural" in Greek, and *lineatus* refers to the lines found on the beetle's elytra.

OF BEETLES AND MEN

The Elateridae comprised about 200 species in France. Some Elateridae display sparkling colors, such as *Ampedus* (below, to the left), which develop in hollowed trees and bear vermillion-red elytra. The Harvest Black-lined Click Beetle, despite its dull markings, is very well known, since it attacks our food. Indeed, the click beetle's larva is as crazy about potatoes as we are, and its appetite for wheat or corn roots is often the cause for significant crop destruction. Three other species of this genus, *Agriotes sputator*, *A. obscures* and *A. sorditus*, are equally adept at devouring our crops.

Red beetle of the genus Ampedus

FOUR YEARS SPENT UNDERGROUND

Click beetles first appear in March. They breed in May and egg-laying occurs until the beginning of July. The female spreads her eggs on soil and they hatch four weeks later. The larvae will go through a very long underground existence (about four years), feeding on roots of both cultivated and wild plants. These larvae, quite pale when young, recognize each other quite easily: they are nicknamed "iron earthworms" because they are hard, cylindrical and slender.

HARVEST TIME

At the end of their growth phase, if they have not been devoured by click beetles, shrews, tiger beetles or birds, the larvae will become nymphs in July and then go through their ultimate transformation into adults at the end of the summer. Most of them will go through their fifth winter in their nymphal nest waiting for spring, to be able to lead their short adult life and then die in September.

Harvest click beetles are infrequent flyers. We sometimes see them on ombelliferous plants or trees in bloom, though they most often live at ground level, their dull coloring guaranteeing the perfect camouflage.

THE ACROBATIC COLEOPTERA

The click beetle's legs are much too short for it to right itself when it finds itself on its back. It does, however, deal with such challenges. It fakes being dead, then lifts its head backward to stretch an elastic muscle. Between its front legs there is a projection capable of retracting within a cavity between the midlegs, and become blocked on its edge – much like a gun's recoil. The accumulated tension in the muscle is freed in one action, propelling the click beetle in the air with a small clicking sound. If needed, it repeats this action, until it finally rights itself.

ACTUAL SIZE
Agriotes
lineatus
(Linnaeus, 1767)
Size: 7–10 mm

Oak Marble Gall Wasp

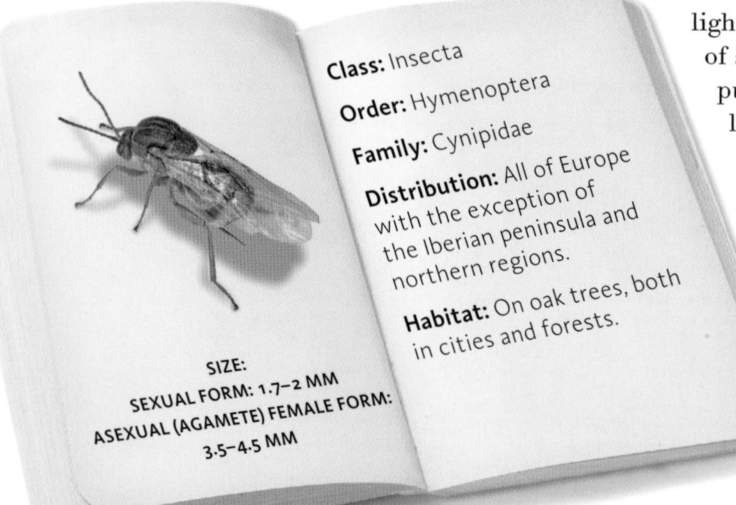

ORIGIN OF THE NAME

The family name Cynipidae (wasps) is derived from *kunos* for "dog" and *ip* for woodworm. The genus name *Andricus* is derived from the Greek *andrikos*, which means "male" or "virile" (a reference to the sexually suggestive shape of the tree galls) while the species name *kollari* is a dedication to Ján Kollár (1793–1852), a Slovakian poet, ideologue and scientist. The insect is better known by the gall it produces, called a wood gall or oak marble gall.

Round Galls

BALLS GROWING ON TREES

A perfect sphere, ranging from 0.6 to 0.8 inches (15 to 20 mm) in diameter, the oak marble gall is essentially a wooden ball. It appears on the pedunculate oak (*Quercus pedunculatus*) in the month of May on a new spring branch. The galls, or "cecidae," are outgrowths that develop from plant material, under the influence of a small hymenopteran: each gall harbors a solitary larva. The marble gall of the oak tree is at first green, then matures, changing to a light brown in August. The adult insect emerges at the end of summer and goes on to bore a tunnel through the tree pulp with its mandibles; it then cracks the external shell, leaving a characteristic small orifice (image at bottom).

A TINY ANIMAL WITH A COMPLEX LIFE CYCLE

These little cynipids (wasps) do not have an easy life cycle: the adults issued from the pedunculate oak's round galls are, in fact, all females; they comprise a generation of asexual, or "agamate" (literally "without marriage"), insects, capable of reproduction without the help of males. During their short life cycle (one to two weeks), they lay eggs (fertile, though not fertilized) on the budding ends of branches of another oak tree, the turkey oak (*Quercus cerris*). The result is the formation of minuscule winter galls (2 to 3 mm long), which develop between the bud scales of the tree. In April to May of the following year, adult males and females appear, smaller than their agamate form. This time, the males fertilize the females, which then begin to search for the right species of oak trees to lay their eggs.

Due to its complexity, the life cycle of this fairly well known little insect was not fully described until the end of the 19th century. Up until 1902, only the agamate form, responsible for the Round Summer Gall, was identified as *A. kollari*. The sexual generation was known under the name of *Andricus circulans*!

The larva (to the left) and adult emerge (below)

YOU DID SAY GALL?

The word "gall" has no relationship to the French word *gale* – a skin condition induced by a dust mite that bores tunnels under the skin of mammals. The plant galls may be due to bacteria, nematodes or dust mites, but also due to flies and aphids. In Europe, however, the Cynipidae induce a gall more spectacular than any other. Galls cause negligible damage to plants. Among others, oak galls, which are very rich in tannins, are frequently used in the tanning of hides and the dying of fabrics in North Africa and in central Asia.

ACTUAL SIZE

Andricus kollari

(Hartig, 1843)

Size: Sexual form: 1.7–2 mm
Asexual (agamete) female form: 3.5–4.5 mm

Woodlands Dung Beetle

The magnificent ventral "bodywork" of the Geotrupidaes

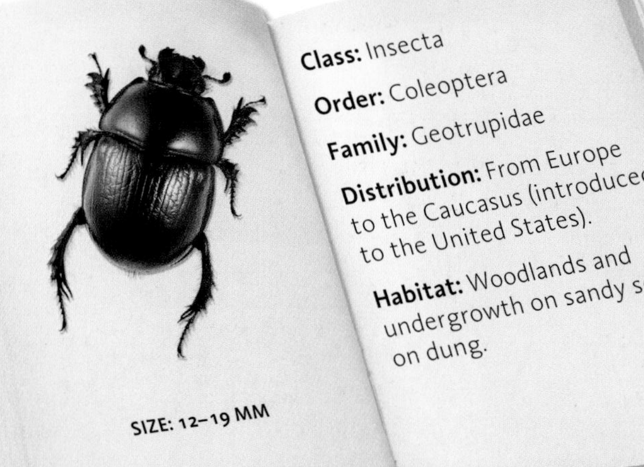

The two sexes cooperate

The dung beetle needs fecal matter to reproduce. It smells the air with its clublike antennae, searching for sources of dung. Once found, it brings the dung back to its nest, which is a dug burrow of horizontal tunnels (sometimes as deep as 16 inches/40 cm). At the bottom of the burrow, the female will deposit one egg in a chamber lined with dung. Thus, under this created mound of dung, the female builds their nursery. Through a lot of back-and-forth activity, the female and the male, relieving each other, bury a mass of dung. (The woodlands Geotrupidae have an affinity for horse manure.) At hatching, the larva has therefore all the nourishment it needs for its initial stages of development. It spends all winter growing slowly in this "pudding" while feeding on it. The nymph stage occurs in situ in May to June; the adult emerges in June to July, its active lifestyle continuing for most of the following year.

Origin of the name

The Geotrupidae (a Greek etymological name meaning that which "digs the soil") belong to the large family of dung beetles. The species name *stercorosus* ("of dung") points to the behavior of this remarkable coleopteran. The genus name *Anoplotrupes* means "that which digs without weapon" (*anoplos*), since the edge of the burrowing legs is effectively less toothlike than in those of the dung beetles of the genus *Geotrupidaes*.

Indispensable insects !

We owe much to these "coprophagous" (excrement-eating) insects. Dung beetles consume and bury excrement (including that of humans), thereby playing a role in its elimination. These beetles are therefore indispensable for the balanced functioning of our ecosystems. Just think that one single cow "produces" about a dozen cowpats a day – about the equivalent of 90 pounds (40 kg)! Unlike its cousins, the woodlands Geotrupidae is only partially coprophagous; it also feeds on fresh mushrooms, rotting mushrooms or both. This insect lives mostly in woodlands and undergrowth on sandy soil, where they can be quite numerous depending on the year.

THE DUNG BEETLES TO THE RESCUE IN AUSTRALIA

In 1788, cows and horses were introduced to Australia's territory for breeding purposes. The numerous local dung beetles, adapted to the little dried marsupial droppings, neglected the large dung of these new arrivals. Millions of these droppings accumulated, thereby chocking the pasture lands. In the 1960s, Australians began importing African and European dung beetles selected especially for their huge appetites, and today, the problem has been almost completely resolved. These dung beetles, however, fall victim to certain treatments given to animals to combat intestinal parasites.

Flying through storms

Dung beetles fly freely at dusk and at nighttime, in a clumsy and roaring flight pattern, especially during warm and stormy days. In the fall, they are occasionally attracted to light. It is during this time that they fall prey to nocturnal vultures (witness their abandoned dung heaps), as well as to foxes and badgers.

An adult on a dung heap

ACTUAL SIZE

91. *Anoplotrupes stercorosus*

(Scriba, 1791)

Size: 12–19 mm

Anthaxia hungarica

Metallic Wood-boring Beetle

The male of the Anthaxia

Larva of Buprestidae

Origin of the name

The name of the genus *Anthaxia* is derived from the Greek, literally meaning "wanting to be a flower" – a lovely image for insects that are so brilliantly colored. The species name, *hungarica*, would lead you to believe that they were originally discovered in Hungary, though paradoxically, this species is not commonly found in this country.

Lovers of yellow flowers

Anthaxia hungarica seeks warmth and is mostly observed in spring and summer in scrubland and dry prairies of the south, in the proximity of oak tree groves. During the warmest hours of the day, the adults leave their shelters to feed on flower petals or stamens, or to mate. They particularly appear to prefer yellow asters without overlooking the flowers of rosaceous plants, including cultivated roses, wild roses and bramble bushes. Observing these insects is made difficult by their quickness – accelerated by the heat – and their ability to fake being dead by falling down into the vast vegetation.

Amazing tunnels!

The legless, flat larvae are characterized by having an extremely large prothorax that gives their body the shape of a construction nail. They build sinuous, odd tunnels both large and flattened, under the trunk or at the heart of branches in pubescent oak (*Quercus pubescens*), green oak trees (*Q.ilex*) or in evergreen oak (*Q.coccifera*). Despite their diet, they are rarely considered nuisances since their preferred spots for egg-laying are in already weakened or sick tree. They also participate in the decomposition of dead wood. After one or two years of larval development (depending on the quality of the wood), the adults spend a winter in their nymphal enclosure before emerging.

A patient larva

Following four to nine days of incubation, a minuscule "worm" gnaws at the interior of its enclosing capsule. It becomes a large larva that, after one month, extricates itself from isolation by a hole that it drilled into the shell with its mandibles. The hole is the same size as its rigid head, and smaller than its chubby body rings (thankfully, the body is flexible). The larva then buries itself to a depth of 4 to 6 inches (10 to 15 cm) in order to build a nest where it will hibernate for two to three winters. The adults emerge at the end of summer to lay their eggs in the fall. It's life cycle requires two to four years to be completed.

Nature's jewels

The Buprestidae abound in tropical regions, though they are also represented in Europe by a few species of modest size. Most of these species are Mediterranean, while others are endemic to northern regions.

Many species display a sparkling external metallic appearance that gave them the bygone name of "millionaires." A little jewel, the brilliant Anthaxia (*Anthaxia candens*), is one noteworthy example of a rare species that is not especially meridian. It is often found sunbathing its elytra (a favorite pastime of this type of insect). Its larva, which values fruit trees, is found mainly under the bark of cherry trees.

THE EMERALD MALE AND THE RUBY FEMALE

In the Hungarian Anthaxia, the two sexes are two different colors. This sexual dimorphism (the morphological differences between the males and females of the same species) plays a significant evolutionary role. Indeed, an individual that has managed to escape predation despite its brilliant colors thus constitutes good parenting stock. This phenomenon is referred to as sexual selection. In arthropods, it is more common to find males to be more brilliantly colored than their female counterparts. The Anthaxia is thus an exception!

Two Mating Anthaxias

Anthaxia hungarica

(Scopoli, 1772)

Size: 7–15 mm

Western Honey Bee

ORIGIN OF THE NAME

Apis is the Latin name for "bees" and *mellifera* means "carrier of honey." Linnaeus, realizing his mistake, changed the name to *mellifica* ("which makes honey") but the International Commission on Zoological Nomenclature, as is customary, retained the first name. The French name is the actual form of a name borrowed from the 13th century – that of the ancient provincial (*abelha*), itself derived from the Latin *apicula*, diminutive of *apis*. Originating in the tropics, bees were "domesticated" and spread via breeding throughout the entire world. When it comes to the simple collecting of wild honey, this habit dates back to ancient times.

A VERY STRUCTURED SOCIETY

The social structure of bees is remarkably stable and quite elaborate. It is comprised of a queen, the only fertile female of the colony, several workers (sterile females) and a few ephemeral males (drones). The queen is essentially a permanent fixture of a colony, and new colonies form by swarming. After a "wedding flight," during which she is inseminated by several males, the queen begins to lay her eggs. The workers assume all other roles in the colony: cleaning the comb cells in the hive, feeding the larvae, fabricating the hive, housecleaning (ridding the hive of any dirt or intruders, including those males no longer considered useful), and especially gathering the pollen from flowers to bring it and nectar back to the hive. To obtain a queen, the workers construct a special cell where the one larva is nourished with an enriched food, called royal jelly.

From left to right: egg, larva and nymph

HONEY, ROYAL JELLY, WAX, PROPOLIS...

We owe a large debt of gratitude to bees for all the good things these hard workers secrete for their larvae. Building from the sugary substances gathered from flowers, honey is, above all, a food reserve for this insect. Manufactured in the crop of the worker bee, honey (partially digested) is passed back and forth from the "gatherer" to the "receiver" to finally reach the right consistency with the necessary amount of enzymes. By a clever ventilation system, the honey is then desiccated in the comb cells of the hive. The royal jelly, of unique composition, enables the feeding of larvae destined to become queens. Beeswax serves to construct the comb cells of the hive, while propolis, a type of tree resin with antibacterial properties, is used as sealant.

S.O.S — OUR THREATENED BEES

For about 10 years now, the collapse of bee colonies has been a worldwide phenomenon. Viruses, fungi, Asiatic hornets, dust mites and bacteria are thought to be the contributing culprits. This phenomenon is worrisome since bees contribute largely to the pollination of plants, including those cultivated by humans. Their ecological and economical roles are truly colossal! Without bees, how would humans feed themselves?

THE LANGUAGE OF BEES

We owe it to Karl von Frisch for the discovery, in 1944, of the famous "language" of bees. Using special dances, bees communicate among themselves and indicate the direction and distance from their sources of nectar. For example, the distance from the source is indicated by the speed at which the messenger executes the dance – the closer the source, the quicker the dance. Generally, the other bees follow the dancer by identifying the smell of the flowering plant on it.

The queen bee (upper left)

Apis mellifera
Linnaeus, 1758
Size: Queen: 15–18 mm
Worker: 11–13 mm
Male: 13–16 mm
ACTUAL SIZE

Araneus diadematus

Diadem Garden Spider

Sad ending for a cricket

Origin of the name

In Latin, its scientific name means, "spider that carries a diadem." Its ancient scientific name is *Epeira*, probably a construct of the Greek epi -, "above" and eirô-, "to bind." The Diadem Garden Spider is also called "cross-bearing spider" in reference to the distinct cross-shape (or diadem) structure on the back of its abdomen.

Tell me the secret of your weaving...

The Diadem Garden Spider is commonly found in a variety of biotopes, as long as there are trees or bushes. At about 3 feet (1 meter) from the ground, it weaves a web called an "orbitel" (i.e., in a regular circle), ranging from 15 to 20 inches (40 to 50 cm) in diameter, comprised of 18 to 42 rays or spokes. It never repairs its web: if the web is damaged, the spider will completely rebuild it, often many times a day if necessary. To remain "economical," it will recycle the silk of the old web by consuming it.

The shape of the webs is characteristic of each genus or family of spiders. Thus, the Giant House Spider (p. 224) construct layered webs, whereas some other spiders construct geometrically-shaped webs with very few rays or spokes, while others keep a hole in the middle of the web.

Natural history...

Adults are only seen from the beginning of the month of August. They patiently wait at the edge of their web and feed on insects that become trapped there. If startled, they will hide under a leaf and drop a long thread to the ground, which they will use to climb down and then they disappear into the vegetation. Males and females will mate at the end of summertime and lay their eggs from September to October. The female can lay eggs many times. She creates a yellow-orangey cocoon, containing several hundred eggs and shielded by a rock or bark, and then she dies. Occasionally, immature females will hibernate in vegetation during a two-year life cycle. Having spent the entire winter shielded, the young ones leave their silk casing the following spring.

...And the supernatural

The Diadem Garden Spiders have been victims of popular superstitions. To lower a fever, or heal a pulmonary infection, old timers would enclose a spider inside two empty walnut-shell halves, glue them together and then require that the sick person wear this around his neck. According to legend, the cross of the Diadem Garden Spider was bestowed upon it at the moment of the Passion of the Christ, in gratitude to the spider, who was said to have woven a protective web over Christ's wounds to keep the flies away!

Diadem Garden Spider cocoon

The male Garden Diadem Spider

I love you, I love you not

By the end of summer, the female has grown to be much bigger than the male. The latter tries to find her for mating purposes. During coupling, he must be extremely careful, since he is at risk of being killed and devoured if the female is not receptive! He loads his pedipalps (in the shape of boxing gloves) with sperm and deposits it into the genital orifice (the epigyne) of the female. The shapes of the pedipalps and the epigynes are two identifying characteristics of each species.

ACTUAL SIZE
Araneus diadematus
Clerck, 1758
Size of Abdomen:
Female: 10–13 mm
Male: 4–8 mm

Arctia caja

Garden Tiger Moth

The caterpillar of the Garden Tiger Moth

WINGSPAN:
MALE: 45–65 MM
FEMALE: 55–75 MM

Class: Insecta

Order: Lepidoptera

Family: Arctiidae

Distribution: From Europe to Japan, as well as in North America.

Habitat: In woodland clearings, fallow lands and gardens.

Originof the name

The genus name originates from the Greek, *arktos*, "bear," a reference to the very hairy caterpillar, which in turn gave rise to the vernacular name of the species: "Bristle-hair Tiger Moth" or "Bear-like Tiger Moth." The species name (from *Caia*, or *Caja*, alluding to a Roman woman of high standing) refers to the beauty of this butterfly's adornments. Its discoverer, Linnaeus, was fond of using feminine names for species with particularly brilliantly colored wings.

Veiled and unveiled

The Tiger Moth is a mostly nocturnal butterfly. In daytime, it can easily pass by unnoticed. Its velvety wings are adorned with cryptic colors in order to blur the animal's appearance, not unlike many large felines. This is why English-speaking people have named it the "Garden Tiger Moth." Its hindwings and its abdomen are of a beautiful red color punctuated by a dark blue-black. When startled, the moth reveals its vibrant colors in the hope of surprising a predator and alerting it to the fact that it is toxic and has a bad taste.

A champion speeding caterpillar

After mating, the female deposits tiny green eggs, similar to miniscule peas, on the underside of a leaf of a nourishing plant. The caterpillars are born in the summer. They do not make many demands on the consumed species and appear to appreciate a number of low-lying plants and several bushes. When startled, they drop to the ground and tuck themselves into little balls like hedgehogs. The difference between them and hedgehogs is that their thick "fur" is pleasantly silken. This will end during their springtime development phase. One surpris-ing fact: when their development is over, the caterpillars literally run to the nymph phase: with remarkable speed, they run on their stomachs to find an ideal location where they can weave their cocoons.

A highly variable ornamentation

This moth does not emerge from its cocoon until July or August. The chocolate patterns on its forewings can spread, to the point of overtaking almost its entire surface; in other cases, the chocolate color regresses to the point that the surface is almost entirely cream-colored. The background color on the hindwings and on the abdomen is red, though it occasionally appears orangey or even entirely yellow. It then becomes difficult to recognize this moth or to distinguish it from the Cream-spot Tiger Moth (see below).

Master of escape

The worst enemy of the Garden Tiger Moth is the bat, which couldn't care less for the coloring-warnings when chasing it during the night. However, our moth is not short of resources. It bears, on both sides of its body, a special auditory organ sensitive to the ultrasounds emitted by its chasing predator. As soon as it senses the signal, the moth's flight pattern becomes erratic and spasmodic, and then it falls to the ground. If that is not sufficient, it then resorts to the ultimate weapon: two "tympani" on the abdomen, enabling it to produce powerful "clicks." These are very close in intensity to those emitted by the bat, which then becomes completely disoriented by this "phantom" echo.

*The Cream-spot Tiger Moth (*Arctia villica*)*

Arctia caja

(Linnaeus, 1758)

Wingspan: Male: 45–65 mm

Female: 55–75 mm

ACTUAL SIZE

Wasp Spider

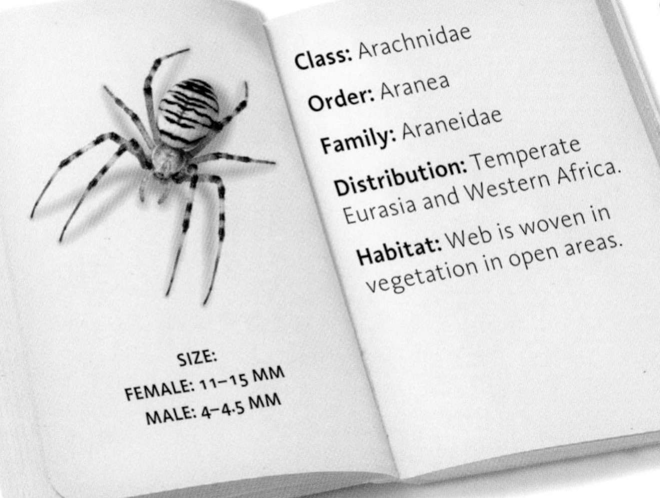

Origin of the name

The Wasp Spider, because of its rayed markings, is also called *Argiope Fasciata* (from the Latin, "marked by bands"). Its genus name, *Argiope* (derived from the Greek and meaning "silver face"), is an allusion to the silver duvet that covers its cephalothorax. The species is dedicated to the Danish naturalist Martin Thrane Brünnich (1737-1827).

Color display

Lying in wait in vegetation, always in very sunny spots, the Wasp Spider awaits its prey, legs spread apart on top of its web. With its brilliant colors, it seems to be the ideal prey itself for birds. No such thing! A predator that has had a bad encounter with a wasp or a spider, will probably remember the color of the culprit. These yellow and black markings become a red alert that protect all animals that mimic the danger warning.

Hunting technique

As with the Diadem Garden Spider (p. 38), the Wasp Spider builds a circular web about one foot (30 cm) in diameter. This web has two large bands of white silk in a zigzag shape in the center. Could this be a visual aid to identify the web and avoid its being destroyed by passing birds? Or could it be an ultraviolet reflector to attract insects? These questions remain unanswered.

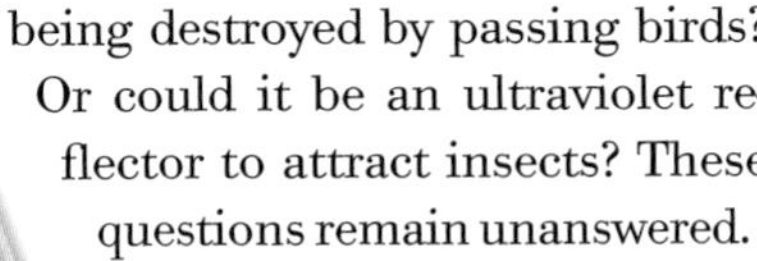

The Wasp Spider perceives its victims' vibrations once they are trapped in the web. Small prey become paralyzed after being bitten and are then carried to the center of the web. The large ones are wrapped in silk, bitten and then injected with paralyzing venom that renders them immobile. Prey that is too large are either ignored or freed from the web.

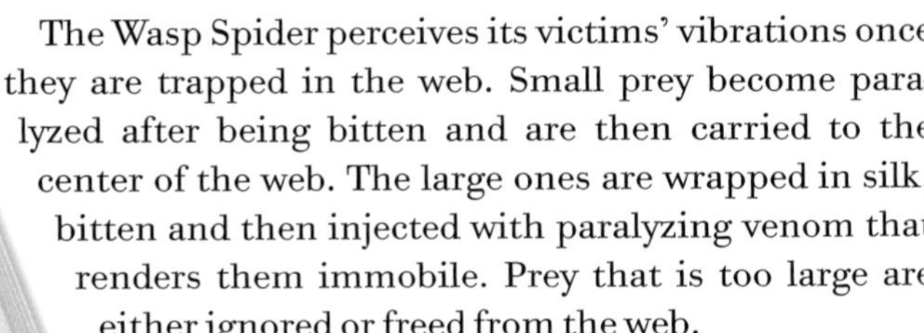

A featherweight boxer

About the end of July, the male, much smaller and less showy than the female, goes about trying to mate. To guarantee a successful reproduction, he must introduce his "boxing gloves" (his pedipalps, already loaded with sperm) into the genital orifice of the female: a dangerous task for such a puny little thing! First, he approaches her by gently tapping on the web so as to make himself known as a suitor. If he is not chased away, or killed, he slides under the female, clasps her and inseminates her. He must then get out of there as quickly as possible. Losing one or two legs during this dramatic encounter is considered the lesser of two evils.

As September nears, the female weaves a very complex cocoon that resembles a brown, upside-down hot-air balloon, in which she will lay hundreds of eggs. The eggs hatch before winter, though the young will spend the winter in the shelter of the cocoon. The mother dies at the first frost. The young leave the cocoon about the end of April, stay together for a while close to it, and then spread out.

SAMURAI SPIDERS

Once a year and for more than 400 years, an unusual competition has taken place in the town of Kajiki, on the Japanese island of Kyushu: the battle of the wasp spiders! Two female spiders, especially "trained," are placed on a stick. The spider identified as "the winner" is the one that bites its opponent on its abdomen, immobilizes it in its web, or severs the silk thread with which it is attached. After the championship, the surviving spiders are released to the environment to sustain reproduction.

Wasp Spider babies

The Wasp Spider cocoon

N.° *Argiope bruennichi*

(Scopoli, 1772)

Size:

Female: 11–15 mm

Male: 4–4.5 mm

Argynnis paphia

Silver-washed Fritillary

The female

Origin of the name

The scientific name is a double reference to the Greek deity of love. When Argynnus, beloved wife of Agamemnon, died, he built a temple in her honor dedicated to Aphrodite; *Argynnus* thus became an epithet when referring to Aphrodite. *Paphia* means "from *Pathos*," the city where, after her birth from the sea waves, Aphrodite (Venus) landed. Fabricius, the creator of the genus *Argynnis*, probably also played on the meaning of *argyros*, "silver," in reference to the markings on the lower surface of the wings of these butterflies. The French name "Spanish tobacco," an allusion to the general reddish-brown color of the wings, appears to have been created by two entomologists, Ernst and Engramelle, in their book on butterflies published between 1779 and 1793.

Feeding off violets!

An unusual trait of these butterflies is that the female does not lay her eggs on the caterpillar's host plant, but between the crevices of tree bark about six feet (182 cm) above ground. After hatching, the very young caterpillars devour the shells of their eggs (named chorion) and spend winter in this stage at their place of birth. When they waken to the first warm days of spring, they leave their tree shelters and begin to search for their host plant, the violet (specifically, *Viola riviniana*). Their bodies bristle with ramified spines, the so-called "scoli," that fend off some predators. Towards the end of May, the caterpillars transform themselves into chrysalis, hanging from a stem of their nourishing violet. Two to three weeks later the gorgeous butterfly hatches, after which it becomes traditionally classified among the "pearly" butterflies (a reference to the shiny underside of their wings).

Silver-washed Fritillary at rest

It is quite a success in the forests edgings

In the Silver-washed Fritillary, it is quite easy to distinguish between the "Mister" and the "Madam." The male (on the right) displays dark brown stripes that underscore the longitudinal nervation on the forewings. The female (above), though less colorful, does not have these markings. The above-described stripes consist of special black layers, called androconies , which secrete pheromones. The scent is irresistible to all females! All male butterflies have androconies but their shape differs from species to species: they can be modified layers as in this species, or simply odiferous tuffs of hair on one of the body parts (abdomen, legs or wings). In the Flat-backed Millipede (p. 202), the androconiacal layers are spread over the wings.

A big success at the forest's edge

The Silver-washed Fritillary (or *Spanish Success*, from the French) is essentially a butterfly found in clearings and at the edges of forests, where its caterpillar feeds on the abundant violets found there. It begins to fly mostly in July, though one can encounter individuals up until September. This butterfly prefers the flowers of brambles, from which it gathers pollen as well as from other species of plants that grow on flat plains to altitudes of 5,000 feet (1,524 m). Both sexes prefer to live at the top of trees and only descend to ground level to feed and reproduce when there is plenty of sunshine.

The Silver- washed Fritillary caterpillar

Argynnis paphia

(Linnaeus, 1758)

Wingspan: 70–80 mm

Atypus affinis

Purseweb Spider

In a defensive position

Class: Arachnidae

Order: Araneidae

Family: Atypidae

Distribution: From North Africa all the way up to Denmark; common in France.

Habitat: Forests or moors; prefer calcareous and sandy soils.

The mygalomorphs distinguish themselves from other spiders by the shape and disposition (said to be "orthognathous") of their venomous fangs: vertical and perpendicular, like those of snakes. To sting, the mygalomorph hoists its fangs upward (upper photo). In other spiders, the fangs are horizontal and overlap.

All in all, it ends pretty badly

To make its nest, the female Purseweb Spider digs a burrow or tunnel of about 8 to 16 in (20–40 cm) in depth, lined with a thick sock-like tube of silk blocking the entrance. By fall, the male is searching for a mate. The female signals her presence by impregnating the deposited "sock" with her pheromones. To make himself known, the male beats his front legs and waits. If he is not rebuffed, he pierces the silk sock and enters the tunnel to mate. They spend part of the winter together, but the male eventually ends up devoured by his partner.

After having blocked the entrance to the tunnel, the female spends the winter hibernating (diapauses) at the bottom of her shelter. At the onset of the following summer, she lays a hundred or so eggs in a tiny cocoon that will hatch by the end of summer. The mother tolerates the presence of the young larvae in the family nest one more winter, after which they leave to start their own lives by first building nests. The lifespan continues for four years for males and six years for females. A mature *Atypus* can live another seven years or more.

Origin of the name

For the Greeks, *mugalé* (from *mys*, "mouse," and *gale*, "weasel") designated the shrew, which they considered to be venomous; in 1802, Latreille revised the name for the large velvety spider, calling it "*mygale*." Strangely, over the years, the word "musaraigne" came from the Latin *mus-aranea*, which means "mouse-spider"! The scientific name is a little more difficult to comprehend: *Atypus* ("atypical") refers perhaps to the shape of the thorax, different from other mygalomorphs and *affinis* ("parent" in Latin), is undoubtedly an allusion to the French species *A. piceus*, previously described.

Mygalomorphs in France!

Mygalomorphs are for the most part tropical, often very hairy, and occasionally dangerous, repulsive and disgusting to arachnophobes. So much for clichés. However, about 20 or so species are found in France, with one living in Paris in the *Jardin des Plantes* ("Plant Garden"). French mygalomorphs are fairly small however, about 0.75 in (2 cm) and live buried within burrows dug in the soil.

REAL SILK SLIPPER

Three species of mygalomorphs live in our environment. The *Nemesia* (family Nemesiidae) and the *Cteniza* (Ctenizidae) are uniquely southern. They lie in wait for their prey a tunnel closed with a little silk, hinged trapdoor. The *Atypus* have a different technique: they build a silk "sock" about 4 in (10 cm) in length (opposite), perfectly invisible on the surface of the ground because it is covered with plant debris and soil. When a prey comes in contact with it, the resultant vibrations alert the hidden spider, which slowly leaves its refuge, invisible under its silk. It kills its prey by piercing it with chelicerae through the silk, and then cuts through the sock to recover its little snack and carry it to the bottom of its tunnel.

Atypus affinis

Eichwald, 1830

Size:

Male:

10–12 mm

Female:

12–22 mm

Oak Tree Apple Gall Wasp

Class: Insecta
Order: Hymenoptera
Family: Cynipidae
Distribution: Europe, Western Asia and North Africa.
Habitat: On oak trees.

SIZE:
WINGED FEMALE: 2.5–3.2 MM
WINGED MALE: 1.9–3 MM
AGAMETE FEMALES: 3.5–7 MM

Gall on oak tree

Origin of the name

The name *Biorhiza* originates from the Greek *bios*, "life," and *rhiza*, "root," in reference to the two egg-laying sites. The species name originates from the Latin, *pallida* "pale." We call the gall made by this insect "oak apple" – lovely imagery!

The home and the shelter

The cynipidae have the ability to manipulate the genetic makeup of the cells of certain plant structures. These little hymenopterans thus construct, via their host plants, a structure destined to protect and nourish them during their own development! This is not just any archaic form of tumor. On the contrary, the cells become organized as tissues that perform different functions (protection, habitat, feeding of larvae). These constructions are rarely abundant enough to cause severe damage to the plant. Please do not destroy them – the shapes and colors of these gall will embellish your garden.

Parasitic Wasp of Cynipidea

Act I: in the buds

The Oak Tree Apple Gall Wasps develop on different species of oak. In the cold of winter, a wingless female of *Biorhiza* (below, left) lays more than a hundred eggs in one bud. Bit by bit, a spongy gall develops, often reaching 2 in (5 cm) in diameter. Inside the gall, compartments are formed wherein larvae develop. Their growth reaches its peak in July, at which point they change into nymphs. A short time later, the adults emerge by digging tunnels in the gall's pulp, covering its surface with tiny exit holes. Depending on the case, the "apple" produces uniquely males, uniquely winged females or representatives from both sexes. However, many larvae never develop because they are devoured by the larvae of another miniscule Hymenopteran (see above illustration), that lays its eggs in the same gall.

Act II: in the roots

The females (winged this time – at right) are fertilized by the males. They then bury themselves in the soil in search of the oak's root system, where they, in turn, will lay their eggs. The gall that develop resemble root nodules, often gathered in grape-like bundles. The larvae will take two to three years to fully develop. They will produce only wingless females, which will have to crawl to the top of the oak trees to find the buds into which they can lay their own eggs to form the new "oak tree apples" – and all of this in the heart of winter!

PUZZLE FOR THE NATURALISTS

For quite some time, the two forms of this species have baffled naturalists, to the point that they first described them under different names assigned to each genus and species. Following the first description in 1791, more than a century elapsed before naturalists understood that we were looking at two forms of the same species during a separate phase of its life cycle. We have yet to understand how, from the same genetic stock, we end up with two generations so radically different. This story illustrates well how there remains much work for the 21st- century naturalists to do, even when it comes to common species!

Wingless agamete female on an oak tree bud

Biorhiza pallida

(Olivier, 1791)

Size: Winged female: 2.5–3.2 mm

Winged male: 1.9–3 mm

Asexual female: 3.5–7 mm

Blattella germanica

German Cockroach

The female and its ootheca

The male

Origin of the name

Blattella is the diminutive of *blatta*, the Latin name of this cockroach. Its species name has nothing to do with its origins, but rather stems from the prejudice that states that a invading species can only come from a neighboring country: in fact, this cockroach is called "Prussian" in Russia and "French" or "Dutch" in Germany! We also call it *cafard* (from the Arabic *Kafir*) or *cancrelat* (from the Netherlander *Kakkerlak*), though originally it was first named the Asian Cockroach (*Blatta orientalis*) and secondly, the American Cockroach (*Periplinata Americana*).

The cockroach has ancient origins

Cockroaches are very old: forms very much like those existing today were thriving more than 300 million years ago. At that time the climate was hot and very humid, conditions still in demand by them to this day.

Some of the "domestic" species which invite themselves on a regular basis into our homes (including the ritziest!) all originate from the tropics: making their way here as clandestine passengers, they have found ideal conditions and abundant sources of food.

The most impressive is the Asian Cockroach at 1.5 in (4 cm) in length. Generally, it does not frequent apartments but prefers caves or bakeries. The German Cockroach is distinguished by its two darker bands on its pronotum. There are also several wild species of this cockroach with discreet behavior patterns that contribute to the decomposition of plants.

A festival of aromas

Females give off pheromones to attract males. A ballet-like contact of their antennae follows and the individuals stimulate one another by producing further aphrodisiac substances. The male lifts his wings to reveal glands from which the female licks the secretions. The female mounts her companion who then need only insert his spermatophore into the genital orifice (cloaca) of the female. In the same manner as its cousin, the Praying Mantis (p. 170–71), the cockroach fabricates an ootheca to protect its eggs. The developmental process of this insect is of the hemimetabolic type, meaning that, at hatching, the larvae resemble the adults (smaller and without wings) and they live the same lifestyle.

To live happily, live in hiding

Cockroaches live in groups but, since they are nocturnal and shun light, they are hard to observe except for the fact that they are prolific! The odors they produce are very noxious and the foods they visit are forever spoiled by the contact. Cockroaches do not fly but run at an incredible speed. Thanks to their flattened body shape, they can squirm through the tiniest of places; it is therefore imperative to eradicate them by treating the entire house at once.

Egg case

When only at the miniscule embryonic stage, cockroaches are particularly well taken care of. In fact, the mother keeps her valuable progeny very close by in an ootheca. This is a type hermetically sealed, isolated box or case in which the embryos are lined up while awaiting hatching. The German Cockroach is the only species that carries her ootheca right up until the hatching of the eggs.

Very young embryos

ACTUAL SIZE

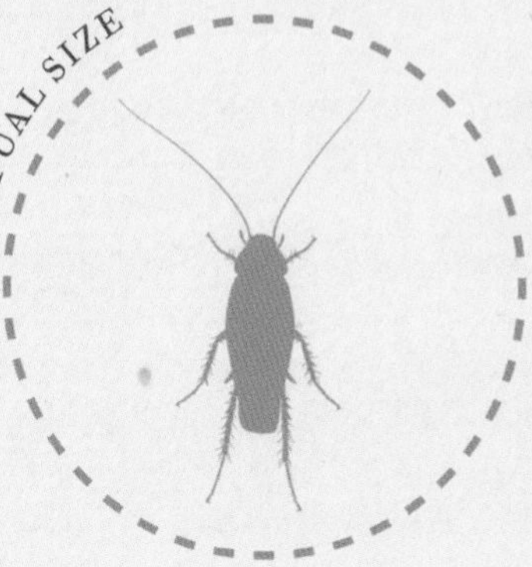

Blattella germanica

(Linnaeus, 1767)

Size: 10–16 mm

Terrestrial Bumblebee

Origin of the name

The Latin *Bombus* designates a quiet noise, a "buzzing or humming" sound; it is a fact that the flight of Bumblebees is not something that passes unnoticed. The French name, *bourdon*, as well as its English counterpart, *bumblebee*, both originated from onomatopoeia, linked to the noise made by this insect. The Bumblebee also carries the common name of "white butt." Unlike their cousins, the wasps and bees, Bumblebees are much less aggressive when attacked, though their sting, which only the females can cause, is equally painful!

The establishment of a colony

Protected by its thick "fur," the Bumblebee is one of the first harbingers of spring. In the first two weeks in March, the young queens that are born that year, protected throughout winter under bushes or in moss at the foot of trees, begin their awakening. Flowers are still rare at this time, though the catkins of willows offer them some source of pollen and nectar. Once sated, they take off in search of an ideal place to establish their future colony, often a small mammal's burrow, ideally suited to protect their progeny from predators and inclement weather conditions. The queen constructs small wax urns, in some of which she stores honey and pollen, while others are destined to receive her eggs. She broods her eggs by working her wing muscles to keep them warm. After three to four days of incubation, the larvae hatch. The mother then dips into the well-stocked urns and feeds her young by regurgitating a mixture of nectar and pollen into their open mouths. The larvae develop in four stages, each interrupted by a molt. At the pre-nymph stage, the young stop feeding and begin to construct a silk cocoon. Young worker Bumblebees will emerge from this cocoon after four to five weeks.

Bumblebee larvae

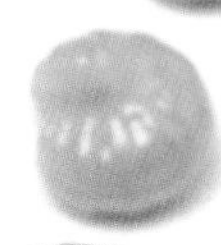

Wedding flight

These first worker Bumblebees, raised by the queen, are small but hyperactive; they will help their mother to raise their young sisters and to enlarge the nest. In fact, other hatchings will follow, giving birth to several successive cohorts of workers that will rapidly expand the initial workforce. Though the workers are quite versatile, the youngest and the smaller "oldest" ones mostly take care of interior tasks, while the older ones and the largest ones forage for food.

At the heart of summer, the colony reaches it apogee. The queen then busies herself producing male offspring (issued from non-fertilized eggs) and sexual females (the future queens), that will soon leave the nest to mate. Around the mid-September, only the hardiest of queens survive, since they have the ability to store fat for their hibernation.

BUMBLEBEES ARE UNSURPASSABLE POLLINATORS

Their opulent fur is a clever trap for pollen, and they can work quickly even in unfavorable weather. Several societies have specialized in the breeding and commercialization of Bumblebees for market gardeners so as to better pollinate their greenhouses.

A Bumblebee colony can have as many as 600 individuals.

N.° Bombus terrestris

(Linnaeus, 1758)

Size: up to 24 mm

ACTUAL SIZE

Bombylius major

Large Bee Fly

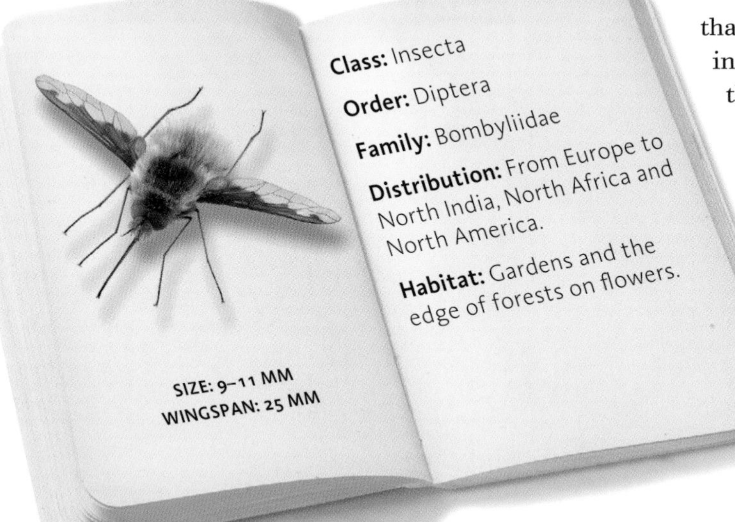

Origin of the name

The scientific name originates from the Greek *bombulios*, "to hum," and from the Latin *major*, "large," in reference to its large size compared to *B. minor* (below), another very common species. Both are also known by the respective names of Big and Small Bee Fly. Though the origin of its vernacular name is largely forgotten, "bichon pet" suits it admirably. It is easy to imagine that in its masquerade as a tiny fluffy toy in a velvet suit, it would have been given the name of that fluffy toy dog, the bichon, much in demand among the aristocracy of the 16th century. The English common name, "Bee Fly," is a much more explicit term.

The deceiver

The Large Bee Fly is distinguished not only by its size but also by the darker coloring on the surface of its wings. With its thick fur, chubby silhouette and attraction to flowers, this insect looks a lot like a hymenopteran, in particular, the true bumblebee. Do not be fooled however! The Large Bee Fly is definitely a dipteran. Despite its long anterior "trumpet," it is not about to sting you. Its long mouth appendage enables it to reach into the deepest flower parts to gather nectar. It can therefore participate actively in the pollination of plants while feeding on their nectar and pollen. It can often be found on violets, primaveras and bugles.

Without breaking the eggs

The larvae (maggots) of the Large Bee Flies live as parasites on the progeny of other insects. Those of the species described herein live on the solitary bees belonging to the families of Colletidae, Andrenidae and Halictidae. Following reproduction, the female will survey the ground's surface to find nests she can easily parasitize. Once the site of egg-laying is chosen, and undoubtedly to protect herself from the sting of the disturbed bees, the Large Bee Fly adopts a unique mode of depositing her eggs. While flying in place, she gathers bits of soil at the end of her abdomen and wraps her eggs in it. This provides strong protection for her progeny which are then thrown, in mid-flight, in the proximity of the opening of the bee's nest. After hatching, the fly larvae reach the bottom of the host tunnel. Taking advantage of the proffered shelter, despite the presence of bees, they feed temporarily on the stored pollen and honey, and then attack the bee larvae before turning into nymphs.

THE HUMMINGBIRD FLY

In the first warm days of spring, you can entertain yourself by observing the agile flight pattern of this insect. The Large Bee Fly, much like the Marmalade Fly (a hoverfly) practices stationary flying. Like the hummingbird, it can feed on a flower's nectar without having to land on it. The rapid beating of their wings is such that it becomes nearly impossible to distinguish individual wings. You may come upon a female looking for a site to lay her eggs, hovering less than half an inch (1 cm) above the ground. If disturbed, she will interrupt her stationary flight by very rapid movements. It goes without saying that the stabilizers (second pair of modified wings) enable Dipterans to execute these impressive flight patterns.

The Small Bee Fly (Bombylius minor)

Large Bee Fly in action

Bombylius major

Linnaeus, 1758

Size: 9–11 mm

Wingspan: 25 mm

Common Bombardier Beetle

ORIGIN OF THE NAME

The name of the genus *Brachinus*, from the Greek, signifies "shortened," a reference to their short elytra; the species name refers to the ability of these insects to spew a corrosive gas while emitting a short, dry noise ("bombardier" also alludes to the same phenomenon).

TINY SOCIABLE CARABIDAE

The Common Bombardier is a small bicolor Carabidae (beetle) that lives in often numerous colonies under the rocks and roots at the edges of fields and in the detritus that populates the sides of roadways. In northern regions, it prefers quite barren calcareous soils, while in the south, it prefers more humid terrain. It likes warmth and is drawn to the bright lights of the stormy summer weather. One can observe it almost year round, though more likely in springtime, in the company of many other carabidae. It feeds on both plant and animal matter.

SIX-FOOTED CHEMICAL CANON

This peaceable insect is equipped with a secret weapon: if threatened, it can project a burning liquid (212°F/100° C!) from its anus on its aggressor, with deadly precision. The German chemist, Schildknecht, explained this prodigious mechanism in 1961. The Bombardier has two glands that secrete an explosive liquid mixture (hydroquinone and hydrogen peroxide). The mixture is stored in two communicating chambers, in the presence of a reaction inhibitor (without which the insect could easily blow itself up!). When threatened, the insect injects the mixture and a chemical catalyst into its two combustion chambers. The release of oxygen and quinine brings the liquid mixture to the boiling point under pressure; then the liquid is violently expelled with the aid of a "canon" at the back of the body. All of this occurs with a surprisingly loud sound!

LARVAE PARASITIZING OTHER LARVAE

Bombardier Beetles have a particular trait which, though not very spectacular, nevertheless enthralls entomologists. The female of the common Bombardier lays her eggs in soil. As soon as they hatch, the larva goes about finding a coleopteran nymph (such as the Carabidae of the genus *Amara*). The larva bites its victim and, like a miniscule vampire, drinks the hemolymph (the "blood" of insects) flowing out of the wound. The larvae go through two successive molts. At the third larval stage, which is e even more voracious, it devours whatever is left of the hapless nymph before morphing itself into an adult. Other species of Bombardiers colonize waterways and similarly attack nymphs of aquatic coleopterans (dytics and gyrins).

THE BOMBARDIER OF DISCORD

The Bombardier has often been invoked to justify creationist dogma. Essentially, they ask: "How does one explain such a complex evolution by simple natural selection? Step by step, it's impossible!" Researchers have felt the need to explain, point by point, the evolution of the organs that have enabled the bombardiers to produce such an explosion without blowing themselves up in the process. But when it comes to matters of faith, no purely scientific argument is totally convincing.

Squadron of Bombardiers: individuals with a red spot on their backs are of the B. sclopeta *species*

ACTUAL SIZE
Brachinus
crepitans
(Linnaeus, 1758)
Size:
7–10 mm

Damselfly

Class: Insecta
Order: Odonata
Sub-Order: Zygopterae
Family: Calopterygidae
Distribution: Europe, Eastern Asia and Northwest Africa.
Habitat: At the edge of running and limpid waters with bushy riverbanks.

SIZE: 44-48 MM
WINGSPAN: 55-70 MM

Female Calopteryx virgo

Charm and seduction

On warm sunny days, the male of the species spends his days chasing small prey and defending his territory against other suitors. If a female appears, he will half open his wings and lift the hind end of his abdomen to reveal the brightly colored segments of his ventral surface. If the female is seduced, she alights: the male then begins a diabolic flight pattern over his beauty. This charming behavior never fails: the male lands on the back of his female, seizes her head with the help of the crooked end of his abdomen, and lifts himself above her. She then bends the end of her abdomen upward to meet the male's penis and form what is known as the "copulation heart." Once fertilized, the female inserts hundreds of eggs one by one in a line on an aquatic plant. Focused on her task, the female becomes vulnerable to predators at this point, despite her coloration, which is more discreet than that of her male counterpart.

The Calopteryx *larva*

Origin of the name

In Greek, *Calopteryx* means "beautiful wings," and in Latin, *virgo* refers to a young woman. The *Calopteryx* is a "miss" or "damsel," a colloquial term in use for many centuries to designate the Damselflies that were svelte and even puny (whereas the large, more robust damselflies were called "mister")!

Young airborn damsel

We encounter the Damselflies mainly in June and July. The fact that the movement of their wings is not synchronized leads to a rather slow flight pattern, and in a somewhat erratic and zigzag manner. Proportionally, damselflies have the largest eyes of the animal kingdom, occupying 80 percent of their heads! Their vision is therefore excellent in all directions. Their long abdomen, consisting of six segments, is a stabilizer as well as a cooling circuit for the blood that feeds the enormous muscular mass of the thorax. Extremely flexible, this insect can bend and twist itself during mating or egg-laying.

Fearsome larvae

Damselflies spend most of their lifecycle in the larval form and in an aquatic environment. The larva breathes via three branchiae at the end of its abdomen; these function as respiration devices as well as oars and rudder. Following the example of its flying elders, the youngster is an excellent hunter. However, its technique favors patience over fretfulness, concealment over exuberance. Hidden away, it surprises its prey (insect larva or young fish) with a folded version of its jaw armed with a fearsome pair of pincers. After two years and several molts, the larvae leave their aquatic environment and wait till the heat of the day to undergo their last molt and slowly transform. In a few hours, their bodies elongate, their wings unfurl and their splendid coloration makes its appearance. They travel for a short while away from their native river, but then return to the limpid and flowing waters they will prefer as adults.

Dragonfly or damselfly?

The order Odonata comprises the Dragonflies (anisopterans) and the Damselflies (zygopterans). How does one differentiate between the two? At rest, the anisopterans retain their four wings in a flat position, whereas Damselflies retain them as befitting a zygopteran: raised and one against the other. Another distinguishing characteristic: with anisopterans, the eyes almost always touch one another, but zygopterans' eyes never touch.

Portrait of Calopteryx

ACTUAL SIZE
Calopteryx virgo
(Linnaeus, 1758)
Size: 44-48 mm
Wingspan: 55-70 mm

Horse Chestnut Leaf Miner

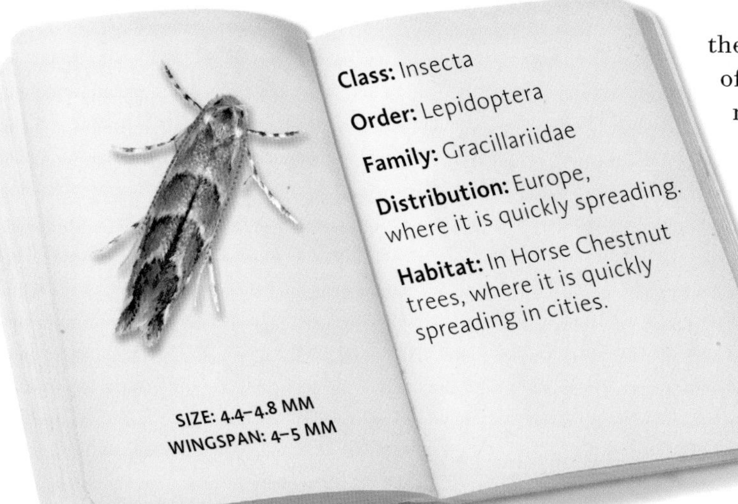

Leaf of an infected chestnut tree

Origin of the name

The genus name refers to the behavior of the caterpillars that dig large tunnels (*camera*, "chamber" or "room" in Latin) in the parenchyma material of leaves. The species name refers to the location (Lake Ohrid in Macedonia), where this little butterfly was first discovered in 1984.

Attacking the horse chestnut leaves

For several years now, it has become apparent in Europe that the horse chestnut's leaves begin to brown in early June and drop prematurely in July. The culprit? The Horse Chestnut Leaf Miner. This small butterfly only creates such havoc at the larval stage, when it digs tunnels in the leaves of the white flowered horse chestnut, resulting in long yellow streaks between the leaf veins. It is unfortunate that this species wreak such havoc, because the adult butterfly is bronzed and nicely colored.

Three generations of miners in one year

At the beginning of May, following hibernation in their litter as nymphs, the butterflies emerge from the dried leaves and gather in large numbers on the trunks of the horse chestnuts, where they breed. The females lay miniscule eggs on the surface of the leaves. Three weeks later, the caterpillars emerge. These enter the leaves and begin to devour the cellular layer just under the epidermis, which is abundant in sugars. After a period of three to five weeks, they metamorphose as butterflies in the leaves. This new generation of butterflies is followed by a third one in September.

Secrets of their success

The rapid spread of this insect is linked to several differing factors which may escape us at first (sadly, the same can be said of species that are quickly disappearing!). It can be attributed to the small stature of the butterfly, which causes it to be overlooked by many predators. Although several parasitic species of hymenopterans have adapted to this invader, the degree of parasitism remains quite weak, which limits their efficacy. The leaf miner's host plant is very abundant in Europe (particularly in cities!) and they have been noted to attack "substitution" trees (maples and other species of horse chestnuts). It has also been noted that its spread is facilitated by long communication routes such as highways, rivers and railways. It is thus quite probable that the Miner will continue to survive for many years to come.

A Lightning-Fast Spread

Making its appearance in Macedonia in 1984, the Horse Chestnut Leaf Miner first colonized central Europe, then Western Europe at an extraordinary speed. It reached France in 1999, England and Spain in 2002, Denmark in 2003 and continues its relentless progression. To this day, nearly all European countries have been affected. In France, only Corsica has been spared – for the time being at least. We do not know the cause of this lightning-fast spread, though one suspects that climate changes have a role to play in it.

The larva of the Horse Chestnut Leaf Miner

Nymph in a leaf

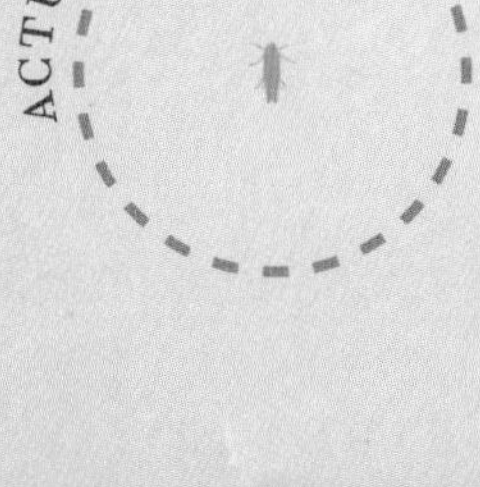

Cameraria ohridella

Deschka & Dimic, 1986

Size: 4.4–4.8 mm

Wingspan: 4–5 mm

ACTUAL SIZE

Golden Ground Beetle

The Gardener and his mate

Carabus *larva devouring a snail*

Class: Insecta

Order: Coleoptera

Family: Carabidaes

Distribution: Europe, as far east as the river Oder.

Habitat: In gardens and fields, and sometimes in the woods.

SIZE: 20–28 MM

Leathery larvae

At the end of spring, the female lays about 20 to 60 eggs in the soil, isolated in small holes, which she has previously dug. After 10 to 15 days of incubation, the small, leathery and very active larvae will hatch, as hungry, if not more so, than their adult counterparts. Towards the end of summer, following two molts, they have grown noticeably. They occupy comfy niches in the soil, where they can turn into nymphs. After this stage, the insect is still soft and whitish. Many days must pass before its housing thickens and its coloration assumes its promised brilliance. The young adult then makes its way to the surface of the soil, where it immediately begins looking for prey. By fall, it is in search of a refuge in which to hibernate, under a large stone for example.

Origin of the name

The genus name Carabus is a version of the Greek word *Karabos*, used by Aristotle in his *History of Animals* to designate the large coleopterans, as well as crayfish. The word has also given us "scarabs" and "crab." *Auratus*, "golden" in Latin, alludes to the insect's brilliant coppery reflections. This beautiful coleopteran was once quite common in all sandy cultivated soils, thus the nickname "Gardener."

A glutton that loves other small garden animals

Equipped with powerful mandibles, the Gardener is a voracious predator of slugs, snails and earthworms. This little beast can engulf about one and a half times its own body weight each day! It is all the more sad that it has suffered from the overuse of pesticides. Common in gardens, meadows and fields, the Gardener is diurnal, whereas most similar species are nocturnal. As with almost all beetles, it does not fly, since its elytra cover its atrophied, membranous wings. The beetles practice a digestion said to be "extra-oral:" they regurgitate a digestive juice on their masticated prey that transforms the prey's tissues into a fine pulp, which is easier for them to digest.

Remarkable longevity

Adult beetles often live for two to three years, an unusually long lifespan in coleopterans. Generally speaking, it is the larva that lives for several years, whereas the imago (the adult) survives for only one season. The Golden Forest Beetle (*C. auronitens*), a beautiful cultivated species (see below) can even reach the saintly age of six years! One must admit that beetles have a solid defense mechanism against their predators: they spray their aggressor with a shiny and noxious substance secreted by specialized anal glands. This particular behavior has also given the Gardener the other common name of "vinegar bottle."

THE PASSION FOR BEETLES

The beetles, due to their unbelievable diversity in coloration, shapes and designs on their elytra (dimples, striations, ribbings, punctuations, etc.), are among the most collected and studied of all insects. Some forests give shelter to spectacular varieties and many local codes protect them against human capture. Recently, the presence of a unique type of *Carabus auronitens* stopped the planned destruction of a small forest, in which the days will now forever remain calm and pleasant.

Carabus auronitens

Carabus auratus

Linnaeus, 1761

Size: 20–28 mm

CARABUS INTRICATUS

Blue Ground Beetle

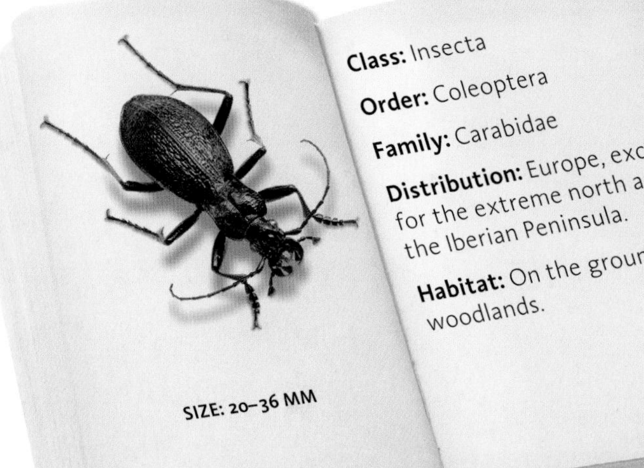

ORIGIN OF THE NAME

The name *Carabus*, proposed by Linnaeus in 1758, is drawn from a name for coleopterans first utilized by Aristotle in his *History of Animals*, although we are not quite sure which insects he was designating. The term *intricatus* means "entangled," in reference to the sculpted markings on their elytra.

SPLENDID PREDATOR

This large dark blue-violet beetle displays rough-surfaced elytra laid out in irregular alignments. As with the Gardener or Golden Ground Beetle (preceding page), its long, sharp mandibles allow it to transport its prey – earthworms, and especially slugs, which it consumes in large quantities. Its rather narrow pronotum, its flattened elytra and its long, svelte legs outline a more elegant profile than the Gardener Beetle (previously described). The male is distinguished by its much larger tarsi than those of the female (which is the case for all beetles and for many other coleopterans). This species can't fly but can scamper very quickly over short distances.

WINTER IN TREE STUMPS, SUMMER IN THE RAIN

As with many beetles, this one is active at night and prefers moving about after a fresh rain when the prey is most abundant. In winter, it goes through a long diapause in worm-eaten tree trunks or in sandy soil at the base of fallen trees.

A DECLINING SPECIES

The Blue Ground Beetle is fond of humid old woodlands that are populated with oak and beech, with abundant supplies of dead trees. Since it can't maintain its population in woodlands where there is intensive and systematic exploitation, this beetle is becoming rare. The species has been tagged with the "vulnerable" status on the red list of endangered species since 1986. Threatened in Belgium, the Netherlands and Denmark, it is barely surviving in southern Great Britain, where it never seems to have been very common. In France, the Blue Ground Beetle is still relatively widespread. In the Île de France, it is limited to the Fontainebleau and Compiègne woodlands, where it is commonly found locally.

Sad end for an earthworm

Eggs of the Blue Ground Beetle

THE COLORATION OF COLEOPTERANS

While birds are renowned for their brightly colored plumage, insects do not pale in comparison! In coleopterans, the origin of the bright colors is not due to pigmentation, but is structural. The external carapace's cuticle in insects is a complex material constructed in microstructural layers. The magnificent iridescent colors are the result of decomposition or of the light diffraction pattern created by these microstructures. The surface relief on the insect also plays a role in determining the diffusion patterns of these internal colors. Finely alveolated, the cuticle surface of the Rural Tiger Beetle (p. 86) presents a very different pattern from our beetle or of the superb Calosome (at right).

ACTUAL SIZE
Carabus intricatus
Linnaeus, 1761
Size: 20–36 mm

Cassida rubiginosa

Thistle Tortoise Beetle

Origin of the name

Cassida, "shield" in Latin, refers to the massive form, suggesting a carapace, of the elytra and the pronotum of this insect. Only the antennae and the legs' extend beyond them. This gives the Cassidae a shape not unlike that of miniature tortoises, which in fact has given them one of their common names – "tortoise beetles." The species name *rubiginosa*, "rust coloring," references the brown spots at the base of the elytra and around the scutellum. This is also known as the Bloody-nosed Beetle.

Family portraits

The Cassidae do not resemble most typical Chrysomelidae (Chrysomelae, of which a splendid example is illustrated on p. 80 as well as other Potato Beetles); they constitute a distinctive subfamily (the Cassidinae), close to the Hispinidae (p. 142), which are also Chrysomelidae of unusual shape. *Cassida rubiginosa* is one of the most common species. It is found on all kinds of asters: thistles, centauries or even artichokes. It should not be confused with the green Cassidae (*C. viridis*), which is as common though much larger, and of a soft green color that is difficult to detect on mint plants, its preferred food source.

A protective parasol over the larvae

The reproductive cycle extends over much of springtime and summer; the large, flattened shape of the carapace of these insects does not easily favor coupling. Once fertilized, the female lays three to six eggs grouped together in a sort of shell on the host plant. She then covers them with excrement: one strange way to keep predators at bay!

From these eggs will hatch green larvae spiked at the rear with forked and spiny appendages. Covered with dried excrement and old skins, these appendages can fold above the animal not unlike a parasol or umbrella. It appears that this odd behavior protects the larvae both from dehydration and predators. The larvae, as do the adults, feed off the parenchyma (internal cellular material of a leaf) tissue of their favorite host plants. After about three weeks, the larvae affix themselves to a leaf, and remain there for six to 10 days until they pupate into nymphs. The nymph shape is spectacular (see photo below). The adults fly rarely and move about their host plants by holding on with suction-like pads equipped with lubricated hairs on their "feet."

Larva of the Thistle Tortoise Bee

The extraordinary Thistle Tortoise

PERFECT MIMICRY

Because of its morphology, one must seize the animal and turn it over in order to see a bit more clearly what it best looks like. However, capturing them is not so easy, since its color blends very easily with its host plant. On top of this, the beetle, upset over this intrusion, quickly folds in its legs and falls to the ground to "drown" in the vegetation. It then becomes impossible to find.

Rust-spotted Thistle Tortoise Beetle: a mini-attack vehicle

ACTUAL SIZE

N.° Cassida rubiginosa

O.F. Müller, 1776

Size: 6–7 mm

CENTROTUS CORNUTUS

Treehopper

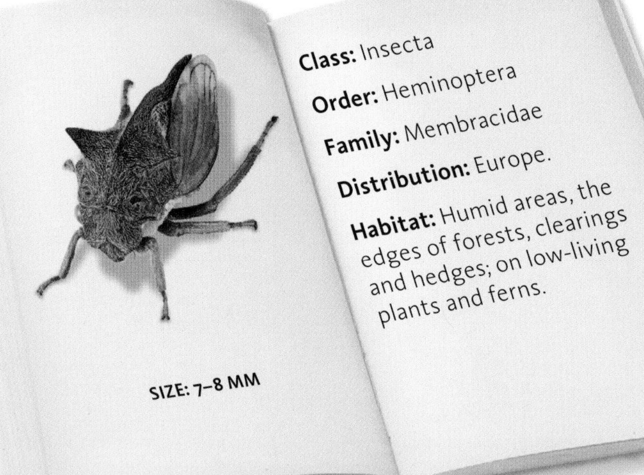

An impressive face

ORIGIN OF THE NAME

The genus name *Centrotus*, from the Greek *kentron*, "thorn" and *otos*, "ear," makes allusion to the lateral horns on the thorax. The species name *cornutus* ("horned") refers to the same structures. In French it is known as *Demi-Diable*, "Semi-Devil," and often is commonly known as *Punaise à tête de chat*, "Bug with a cat's head."

THE HORNY BEAST

The Semi-Devil is one of the rare European representatives of the Membracidae family, widespread in the tropics (particularly in South America). It is encountered locally on low-living plants and ferns, in humid regions, at the edge of forests, clearings, poplar growths, etc. It is easily recognized by its odd "helmet" (a pronotum modification of the first segment of the thorax) embellished with two angular lateral horns and a very long median sinuous expansion that extends over the wings. Adults easily fly and jump from one spot to another, although juveniles lack this ability. The female has an ovipositor to enable her to lay her eggs in the stems of plants at the end of summer. The eggs hibernate and the juveniles develop in springtime and then become adults (there is only one generation per year).

A PRICKLY QUESTION

What role can we assign to the horny protrusions on these beetles? In some instances, the answer seems clear: certain species so closely mimic a plant's spine, both in shape and in color, that when they are lined up on a stem, the total effect is of a thorny tree branch. It is even more difficult to assign a role to the extraordinary appendages (halters, sinuous or ramified horns, spiny balls, hooks, fan-shaped comb) present in some tropical species. Maybe their sole role is to render their owners all the more difficult to swallow?

OTHER LITTLE DEVILS

The Semi-Devil has a cousin in the south in the Little-horned Devil (*Centrotus chloroticus*), which is smaller and more transparent. Also close in shape, the Little Devil (*Gargara genistae*) distinguishes itself by with a median horn and shorter, lateral growths. It can be found, flying and jumping vigorously, at the edge of new woodlands where brooms (with their yellow flowers) and leguminous plants are an abundant food source for its larvae.

A TRAFFIC-MONITORING BEETLE

The Membracidae Bison (*Stictocephala bisonia*) or Buffalo Treehopper is also a spectacular species (see below). Originating from South America, this species was accidentally introduced to Europe in the 19th century, and its spread was meteoric, since it is now observed throughout that continent. Overhanging the head, the enormous pronotum is equipped, up front, with two spiny horns and bears a median horn-shaped appendage that swoops to the rear. Despite its massive shape, the Buffalo Treehopper can fly and jump very well. It can be encountered in fallow lands and in sunny orchards where it grows on various trees and bushes. This insect may wreck local havoc when a large proliferation occurs in some years. The havoc particularly affects apple trees, where the new bark is damaged during the egg-laying.

Larva of Semi-Devil

The membracide Bison (Stictocephala bisonia)

69
Centrotus
cornutus
(Linnaeus, 1758)
Size: 7–8 mm
ACTUAL SIZE

CERAMBYX CERDO

Great Capricorn Beetle

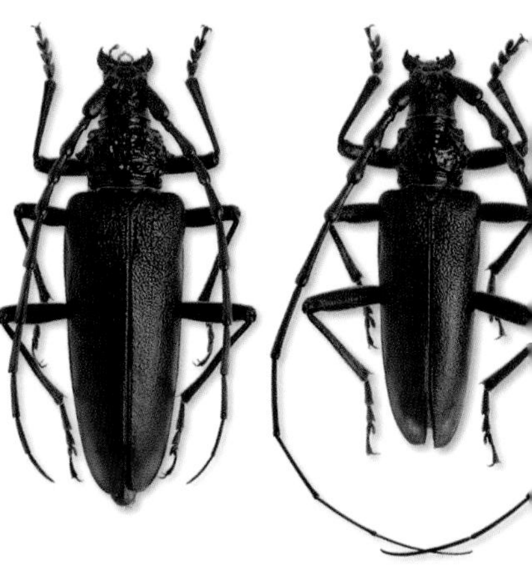

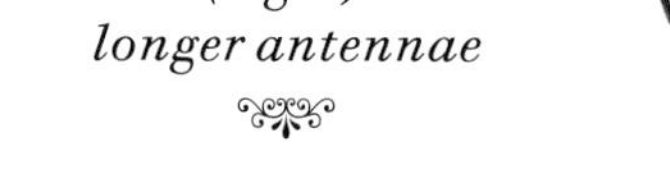

The male (right) has much longer antennae

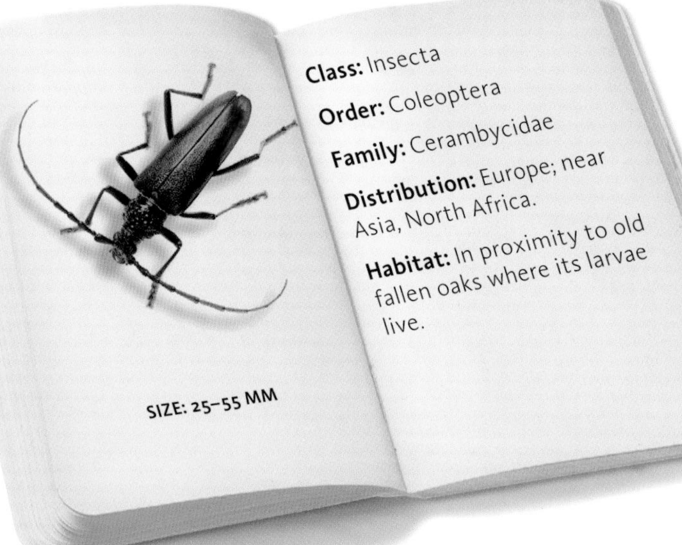

ORIGIN OF THE NAME

The name *Cyrambyx* is of Greek origin and refers to an insect with long "horns"; *cerdo* is a craftsman (undoubtedly a carpenter, if one considers the biology of the insect). "Capricorn" originates from the Latin and means "sheep's horn" – a reference to the characteristic appearance of the very long antennae. Its other common name, from the French *longicorne*, is even more obvious – referring, of course, to longhorn sheep.

THE BIGGEST AND THE MOST BEAUTIFUL

With its impressive antennae (they surpass the body length of males, but are much smaller in females), the Great Capricorn Beetle is one of the largest European coleopterans. Its brilliant black elytra turn reddish brown at their extremities. This beautiful insect flies heavily, with its antennae spread toward the front. When in search of a partner at nightfall, it can be found in forests or parks where old oak trees grow.

GIANT LARVAE

The female lays her eggs in the crevices of old oak trees. The larva, resembling a legless, large white worm, is enormous, reaching lengths of 3 to 3.5 inches (7 to 9 cm) after three to four years. Carrying hardened dorsal plates characteristic of the species, the larva feeds by digging a tunnel in the sapwood and finally reaches the heart of the trunk. The attacked tree does not necessarily die, but its woody tissues are damaged. The larva pupates into a nymph in a niche sealed with a calcareous material, and the adult appears in early autumn, even though it does not reappear again until the following year. Adults can meet one another during the entire summer. They lick the sap that drips from the oaks; they never gather pollen from the flowers.

CAPRICORN, SHOW ME YOUR HORNS...

When it is stressed, the Great Capricorn Beetle sends out a very perceptible, tiny grating sound by rubbing the chirring plates situated between the elytra and the pronotum. No one is permitted to capture this insect in France as it has been a protected species over the entire national territory since July 22 , 1993. This is a laudable measure; however, there is still considerable hypocrisy in the fact that it remains possible – outside of strictly protected forests – to transform the cut old oak tree growth that shelters these insects and transform it into planks of wood.

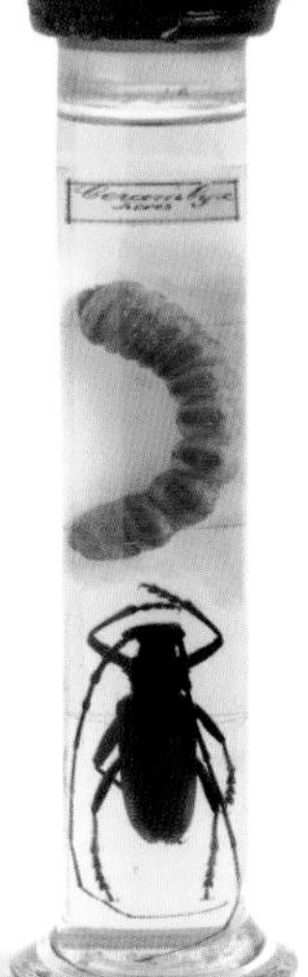

FOSSILIZED TUNNELS

The Great Capricorn Beetle prefers continental old forests with hot summers, and it has never been found in Great Britain. However, by exhuming the trunks of fossilized oaks, buried for more than 10,000 years in the British peat bogs, characteristic tunnels of this coleopteran (a species now extinct on that side of the Channel) were discovered inside the wood.

Great Capricorn Beetle near its tunnel

ACTUAL SIZE

Cerambyx cerdo

Linnaeus, 1758

Size: 25–55 mm

CERCOPIS VULNERATA

Red-and-black Froghopper

Froghopper lar

SIZE: 8–10 MM

Class: Insecta

Order: Heminoptera

Family: Cercopidae

Distribution: Temperate Eurasia.

Habitat: Humid prairies, dry embankments, especially in wooded areas, equally on a variety of plants.

Origin of the name

The froghoppers are, for all intents and purposes, very small cicadas. The ancients were not far from the truth when they named this insect *Cercopis*, which is the Latin form of the Greek name for cicada (*kerkope*). The red markings that decorate the elytra of the Red-and-black Frog-hopper (Blood Cercope) inspired the species name to the scientist who described it: *vulnerata* (which means "wounded" in Latin). In English, we call this beetle a froghopper, whereas the French call it a bloody cercope!

It stings ... but only plants

Froghoppers belong to the order Hemiptera, which also includes cicadas, aphids and mealybugs. All of these insects are equipped with biting mouthparts. Froghoppers are also known as sapsuckers.

Cuckoo spit

Froghopper larvae need to live in humid environments. Different survival strategies are adopted by the different species. In the Red-and-black Frog-hopper, the larvae live underground and feed on plant roots. However, other species prefer to remain on the portion of the host plant that is above ground, and they wrap themselves in a mass of froth known as "cuckoo spit." This is the case with *Philaenus spumarius*. The froth can best be seen during springtime on willow or alder branches, or in grasslands (photo below). The cuckoo, which also appears in springtime, was once thought to be responsible for these frothy masses. Other names, such as "witch saliva" and "frog's drool," testify to the fact that this insect was perceived as somewhat mystical.

Churned "urine"

Since the raw sap that it feeds on is low in nutrients, froghopper larva must ingest it in large quantities. It therefore eliminates a great deal of viscous liquid waste, which can be mistaken for urine. Aware of its potential for dehydration, the larva reuses this liquid waste for protection. Under its abdomen, the larva has a cavity carpeted with secretory glands. The product of these glands, mixed with the "urine" and air, allows the larva to form little bubbles that spread around it with the help of abdominal movements. This pile of bubbles has another benefit: it protects the larva from predators such as spiders or ants.

A froghopper (Philaenus spumarius) *in its "cuckoo spit"*

WORLD JUMPING CHAMPION

The flea has recently lost its title as jumping champion to the froghopper. Froghoppers can leap up to 28 inches (70 cm), compared to the flea's 12 inches (30 cm), with an acceleration 550 times greater than the pull of gravity. This is a record! To accomplish this, it suddenly releases the tension built up in a tendon maintained by an anchoring system in the lower legs – in effect creating a catapult. During the jump, it produces a force equal to 400 times its weight (compared to 300 times for the flea, eight for the cricket and two or three for humans).

Exuviate of the froghopper

Cercopis vulnerata

Rossi, 1807

Size: 8–10 mm

CETONIA AURATA

Golden Cetonia

Cetonia emerging from its shell

SIZE: 14–21 MM

Class: Insecta

Order: Coleoptera

Family: Scarabidae

Distribution: Europe and Northern Asia.

Habitat: On a variety of flowers: roses, umbelliferans, and fruit-bearing trees. Larvae in compost or on decomposing wood.

The Cetonia larva moves around on its back

ORIGIN OF THE NAME

The word *ketonia*, already in use by later-day Greeks, has an obscure origin (perhaps a Gallic name meaning wood, or forest). It was revived by Fabricius for this genus name, without any particular reference. The species name (*aurata*, for "golden") of this magnificent insect is, for its part also uniquely appropriate.

THE ROSE CHAFER

The Golden Cetonia flies in summer and fall during the warm hours of the day. Its flight pattern is clumsy, but efficient. Contrary to other coleopterans, such as the European Cockchafer (a dung beetle; see p. 174) it does not lift its elytra to disengage its wings. The Golden Cetonia seeks out flower pollen and nectar, though it doesn't disregard the oozing sap, nor the honeydew of aphids and the ripe end-of-season fruit. Since it spends most of its life on roses or on the flowers of wild roses, it also bears the nickname of "Rose Chafer."

COMPOST-FRIENDLY LARVAE

Golden Cetonias mate on flowers. The egg laying occurs in a soil cavity, and the young larvae hibernate at their third larval stage. When the weather warms up, they resume their feeding: leaf mold, compost, decomposing trees and dung suit them quite well. Pupating lasts about two months. To pupate, the larva constructs a hard shell with very smooth internal walls, made of compacted earth containing its own excrement. The adults can remain in this shelter for some time to emerge at the onset of the fall season, even if the mating doesn't occur till the following spring. In contrast to the European Cockchafer, the Golden Cetonia's larva is not harmful. On the contrary, it contributes to the decomposition of plant material. To have one in your compost pile is a bonus! It also distinguishes itself from the European Cockchafer by having smaller legs and by its manner of getting around by lying on its back, with its "feet" pointing upward. If it is set on its stomach, it flips over again and resumes moving in this very odd way.

CETONIAS OF ALL COLORS

The Golden Cetonia is typically golden-green or coppery, with white transverse striations on the elytra – this is its most common appearance in northern regions. Things are quite different in southern regions, however, where certain hawthorns can carry "grape clusters" of Cetonias, all a bit different from one another, singly colored or two-toned, purple, violet, blue or black.

A CURE FOR RABIES!

In Russia, the Golden Cetonia was once used as a cure for rabies. It was thought that humankind had found a means to combat this disease by using this insect in the form of a dried and powdery form, as an antidote. Needless to say, the discovery of the anti-rabies vaccination by Pasteur in 1885 remains, to this day, the only efficient way of dealing with this terrible disease in humans. For as soon as the first signs of rabies appear, death is usually a foregone conclusion.

Compare the male (on the right with its ventral furrow) to the female (on the left)

ACTUAL SIZE

Cetonia aurata

(Linnaeus, 1761)

Size: 14–21 mm

Meadow Grasshopper

A male Meadow Grasshopper (with long wings)

Class: Insecta

Order: Orthoptera

Family: Acrididae

Distribution: Europe.

Habitat: In meadows and prairies. Not very picky when it comes to choosing its environment, though it does not appear to like areas that are too dry or too humid.

SIZE:
MALE: 13–16 MM
FEMALE: 17–23 MM

Larvae of Meadow Cricket

ORIGIN OF THE NAME

The genus name *Chorthippus*, from the Greek, literally means "prairie horse." This is probably an allusion to this insect's ability to jump, as well as the shape of its corset, which vaguely recalls a saddle – two traits it shares with all grasshoppers. The word "cricket" is an onomatopoetic word, which imitates the "chant" of many orthopterans.

A FAMILIAR GRASSHOPPER

The Meadow Grasshopper is one of the most common grasshoppers found in our region. Its color is generally variable (green, brown, purplish or rose) but its posterior knee joint (an articulation of hopping legs) is always black. They are found from June to November in meadows or on embankments, at the edge of highways and roads, and even in gardens if the grass is not mowed too short. On the other hand, it is rare in swamplands, or in sandy soils such as dunes. Since its wings are very short (reduced down to mere stumps in the female), its gets around mostly by hopping. Strangely, once in a while, you can find a certain proportion of individuals that are "macropterans" (having normal wings).

THE 8TH PLAGUE OF EGYPT

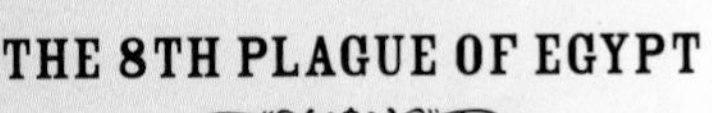

As with all crickets and grasshoppers, the Meadow Grasshopper feeds on leaves of grains, which it chews using its grinding mouthparts. It is a very modest lawn chewer when compared to migratory or gregarious species, such as the Migratory Locust (*Locusta migratoria*) and the Desert Locust (*Schistocerca gregaria*). These veritable scourges of warm regions often appear in groups, occasionally forming clouds of hundreds of millions of individuals that devour entire crops in their wake. The Old Testament even refers to the Desert Locust as one of the Plagues of Egypt!

ACCORDION-SHAPED ABDOMEN

The female abdomen in this species ends with a short egg-laying tube (the ovipositor), which is flexible and in the shape of a beak with four valves (in other grasshoppers it is very long and in the shape of a saber or sword). When laying eggs, its abdomen spreads out like an accordion to force the eggs into the soil. As soon as they hatch, the youngsters greatly resemble their adult counterparts. They grow by successive molts with adult characteristics (wings, reproductive apparatus, and so on) appearing progressively.

TELL ME HOW YOU CHIRP...

Unlike crickets (p. 228), the Meadow Grasshopper chirps by running its "crenellated" femurs, used like a violin's bow, on the reinforced edge of its elytra.

To seduce females, the male not only chirps during daytime but also at sunset. Its chirping is very recognizable: nine linked accents, followed by an accelerated crescendo of two to three seconds, which is repeated every 5 to 15 seconds. The many species of the genus *Chorthippus* are difficult to tell apart without some degree of expertise. However, each one has a different chirp: fast, staccato, whistling and intense for *C. biguttulus*; strident, buzzing and brief for *C. brunneus*; similar to the Meadow Grasshopper, but louder and less hoarse, in *C. palustris* and so on.

The exuviate – what remains after a molt

Chorthippus parallelus

(Zetterstedt, 1821)

Size: Male: 13–16 mm

Female: 17–23 mm

ACTUAL SIZE

Ruby-tailed Wasp

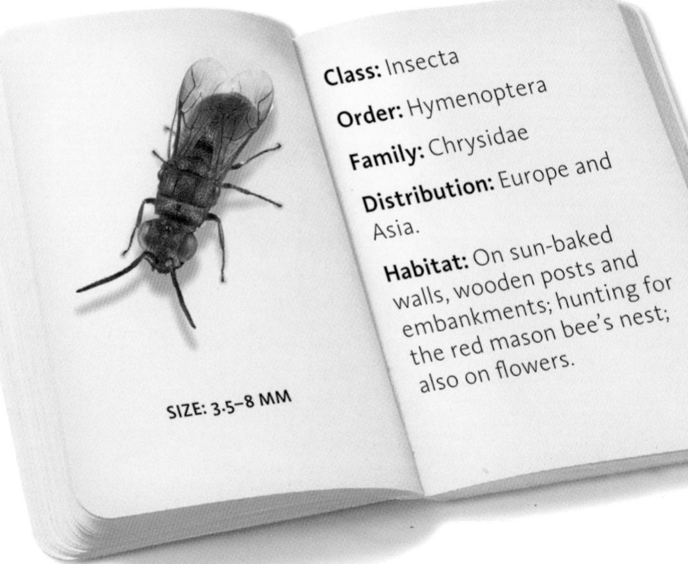

Origin of the name

Chrysis originates from the Greek *chrusos*, which means "golden." The colors of the Ruby-tailed Wasp are very evocative of an open flame – *ignita* means "burning, ardent" in Latin. This insect is also called the Flamboyant Wasp, as well as the Cuckoo Wasp, in reference to its odd growth pattern (see below).

Pleasing to the eyes

We cannot speak of these insects without mentioning their magnificent metallic coloration, in golden yellow, blue, red or green. These colors are not the result of pigmentation but are due to structures at the surface of their cuticle that diffracts light. In the summer, when the sun is at its hottest, you can observe the enflamed *Chrysis* busily gathering pollen, preferably from ombelliferan flowers, or simply perched on rocks. You can also find it nosing about, rapidly flying trying to locate a nesting place of solitary bees, along a wall or a sunny embankment.

The "cuckoo wasp"

Chrysis shares some very dishonest habits with the cuckoo bird. It seeks out the nest of another hymenopteran and uses it for its own offspring. This wasp most often selects the nest of Potter Wasps of the genus *Eumenes*, which resembles small amphorae, or a species of the genus *Odynerus* that makes small tubes out of dirt at the exit of its tunnels. *Chrysis* prefers species that horde their food during the different developmental stages of their larvae. When seeking shelter, it will make sure that the tunnel is already occupied by the host's offspring. It then leaves the tunnel and re-enters, this time backing into the tunnel to deposit its eggs. Its egg-laying tube (ovipositor) is robust and telescoping, with small toothlike structures that enable it to break down the muddy walls and to reach the heart of the tunnel. The female wasp then reseals the hole with a wet mixture of soil and saliva.

The Chrysidae larvae are said to be "parasitoids." As soon as they hatch, they attack not only the food reserves, but also the larvae in the nest. The adult wasps vary in size from small to large, depending on the quantity of food that the larvae will have ingested during their development.

The enflamed Chrysidae is a veritable winged gem

THE ARMADILLO WASP

It is not unusual to see the Cuckoo Wasp being surprised by the solitary wasp whose nest it parasitizes. Though it may not be armed with a powerful stinger, it is not defenseless. It has a formidable armor consisting of a thick layer of cuticle (chitin). Its abdominal segments overlap one another very much like those of an armadillo. When attacked, it folds over into a ball and fakes being dead. The ventral surface of the abdomen is concave, leaving room for it to tuck in its head, antennae and appendages. With the exception of its wings, it leaves nothing to its assailant.

A Chrysidae of the genus Holopyga in a folded defense

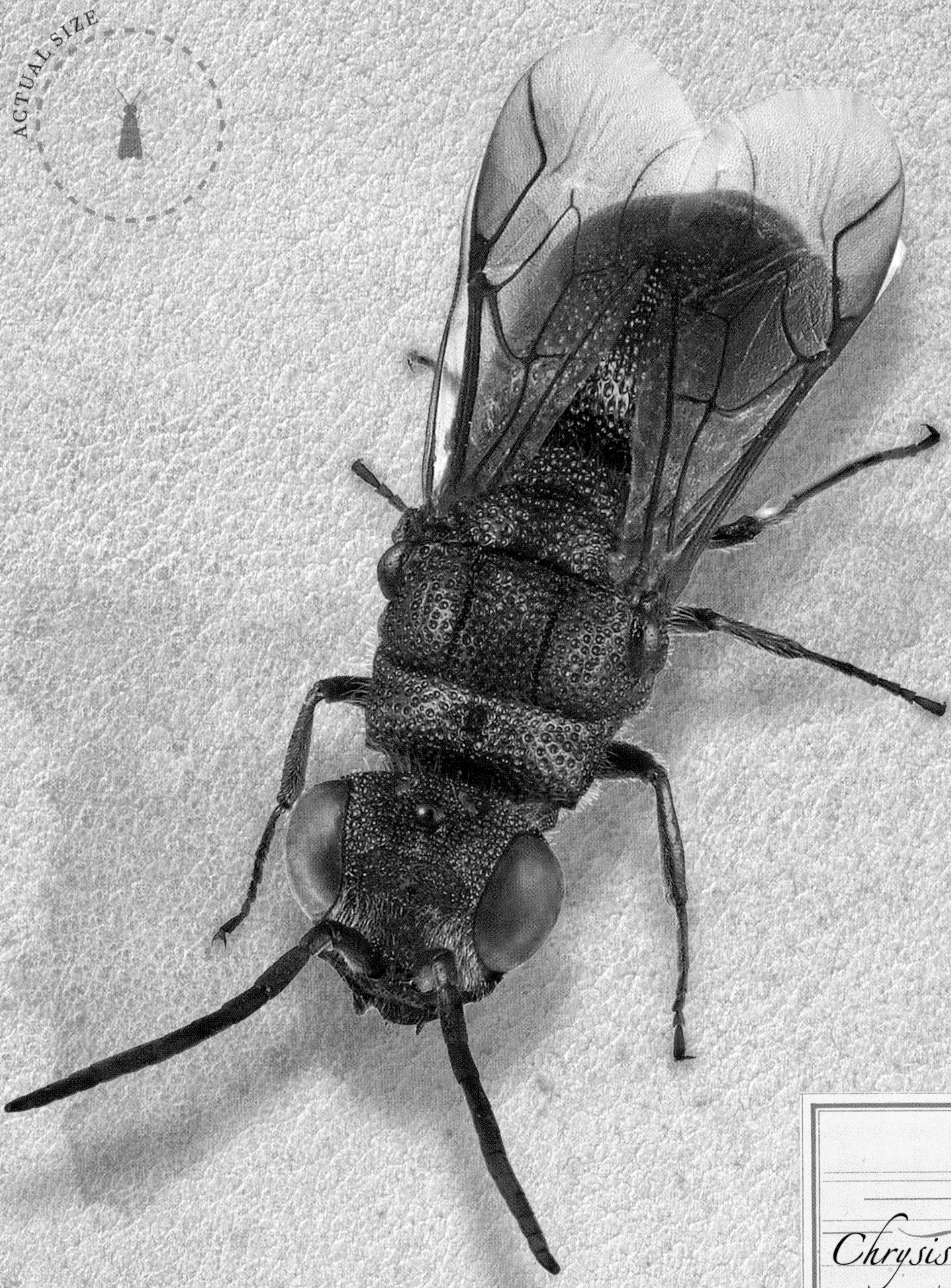

ACTUAL SIZE

Chrysis ignita

(Linnaeus, 1758)

Size: 3.5–8 mm

Chrysolina americana

Rosemary Leaf Beetle

Class: Insecta

Order: Coleoptera

Family: Chrysomelidae

Distribution: Mediterranean Europe and originating in North Africa; found in Great Britain, The Netherlands and Romania.

Habitat: On lavender, rosemary, thyme and other herbs.

Origin of the name

The common and scientific names are derived from the Greek, *chrysos*, which means "gold." This allusion is self-evident since the Chrysomelinae (the subfamily) are striking insects with metallic reflections. *Chrysolina americana* is well known by gardeners under many other names: American Chrysomelinae, or the Rosemary or Lavender Leaf Beetle. Though it is found almost exclusively on lavender and rosemary, the etymological origin of the species name *americana* is quite mysterious since this insect is not found in North America!

The flirt

A veritable little gem, the Rosemary Leaf Beetle decorates lavender or rosemary plants with its round shape and red and metallic blue-green stripes, finely chiseled with pointed lines. It is most commonly found at the extremity of the plants' stalks, on the tender new growths of the previous year. It grazes quietly on lavender and rosemary leaves, which are protected by concentrated essential oils that generally repel most of its predators. The Rosemary Leaf Beetle, though, takes advantage of these essential oils to "perfume" itself, thereby giving its flesh a bitter taste that is noxious to insect-eating birds!

Reasoned cultivations

Chrysolina americana was originally a species of the Mediterranean region. However, its distribution has spread northward for dozens of years, under the combined influence of a warming climate that enables individuals to hibernate in colder regions, and the development of cultivars of lavender and rosemary mostly as ornamental plants.

Its beauty does not necessarily absolve it from its misdeeds, especially those inflicted on cultivated lavender. For most gardeners, the damage is generally marginal with respect to survival of the host plants. On rare occasions, though, this natural balance is disrupted and the beetles cause considerable damage. Simply harvesting the larvae, the adults or both, helps restore a more balanced situation. All that is needed to do this is to place some netting over the plants and shake the stems to release and then destroy these beautiful pests.

DOESN'T AVOID COLD

Astonishingly, the Rosemary Leaf Beetle does not hibernate. With the advent of cold weather, its activities and numbers are somewhat reduced, though you can find this insect in every season. The eggs are laid at the beginning of fall on the host plant, under the leaves or on the stem, and development continues in winter. The larvae remain and feed on the host plant. To pupate, the larvae fall to the ground and bury themselves for protection. Come springtime, the adults emerge from the soil, feed on the plant and continue the life cycle. During summertime, the beetle seems to have disappeared (this is referred to estivation rather than hibernation) and seeks humidity and freshness while becoming dormant.

Chrysomelinae eggs and neonatal larvae

Chrysolina americana

Linnaeus, 1758

Size: 6–8 mm

Green Lacewing

The adult emerges (note h
the wings are crumpled

Class: Insecta

Order: Neuroptera

Family: Chrysopidae

Distribution: Mediterranean Europe; the Canary Islands and North Africa.

Habitat: Prefers new vegetation; the larva can decimate colonies of aphids.

SIZE: 10–15 MM
WINGSPAN: 20–25 MM

during the day it rests under foliage. The adult feeds on pollen, nectar or honeydew.

When eggs are high above, they are less vulnerable

The female lacewing lays her eggs on top of leaves or on branches. She first deposits a substance that forms a long filament by lifting her abdomen as high as possible. This filament sets immediately, forming a pedicle nearly half an inch in length and thinner than a strand of hair, into which she lays one egg that remains out of reach from many predators. A hairy larva emerges, armed with large mandibles and destined to become a fearsome predator (see photo below).

Devouring aphids

Despite its different hunting technique, the lacewing larva, which Réaumur described as "the aphid devourer" is similar to that of the doodlebug – the famous "ant lion." It pierces its prey with its long mandibles and sucks up its content. The larva feeds on aphids, thrips, whiteflies and mealybugs – in other words, on everything that infests our plants! Several generations of lacewings are known to overlap each year. Complete metamorphosis, or pupation, occurs in a narrow cocoon of white silk. The Chrysopidae hibernate in the adult state in low-lying vegetation, though they can often be seen in homes or other types of shelters.

Origin of the name

Chrysoperla lucasina (Lacroix) is a distinct species of Green Lacewing. Its other common name, Chrysope, stems from the Greek *chrysos*, meaning "gold" and, *ops*, meaning "eye," and the scientific name *Chrysoperla* ("golden pearl"), refers very specifically to the eyes of these insects. Some people will go as far as to say that these "golden-eyed damselflies" have the most beautiful eyes of the entire animal kingdom. The species name honors the French entomologist (and painter/illustrator) Hippolyte Lucas (1814–1899), who described several insects of Mediterranean regions and Algeria.

Confusion among species

The Chrysopidae are all neuropterans like the doodlebug (p. 116); as with the doodlebug, the wings of Chrysopidae are displayed, at rest, in a closed coverlet formation. Along with *C. carnea* (the common Chrysopidae) and *C. affinis*, *C. lucasina* belongs to a complex of similar species that have been confused for a long time. *C. lucasina*, a bit more Mediterranean than the others, also displays a unique small comma in the neck. It can be found in bush and grasses, while *C. carnea* is more shrub-loving. Mostly active at twilight or else nocturnal, it prefers cool areas, while

An excellent alternate

As with the mealybugs, the Syrphidae or their cousins the Hemirobiidae (brown lacewings), Green Lacewings are used for biological pest control. A single larva can devour 200 to 500 aphids during its lifetime! This method of control is generally preferable: sites are set up (for example, hay stacks and strips of grassland) to attract and maintain natural lacewing populations. Such shelters can be set up in gardens, and strips of lacewing eggs are also available, thereby limiting the use of chemical pesticides.

The perched eggs of Chrysopidae

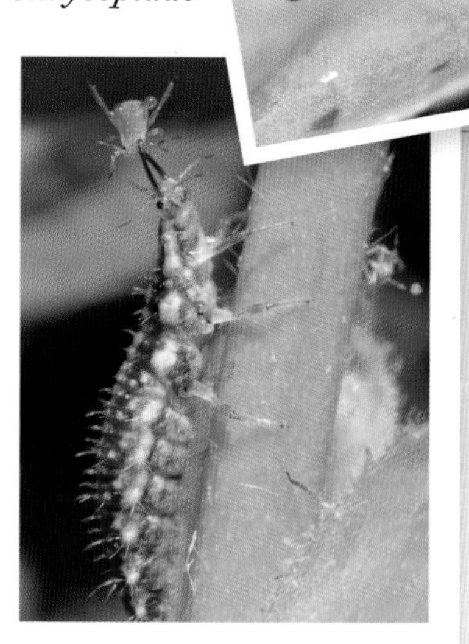

The aphid devourer

ACTUAL SIZE

Chrysoperla lucasina
(Lacroix, 1912)
Size: 10–15 mm
Wingspan: 20–25 mm

Cicada orni

Ash Tree Cicada

Cymbals: the secret of the cicada' chants or chirps

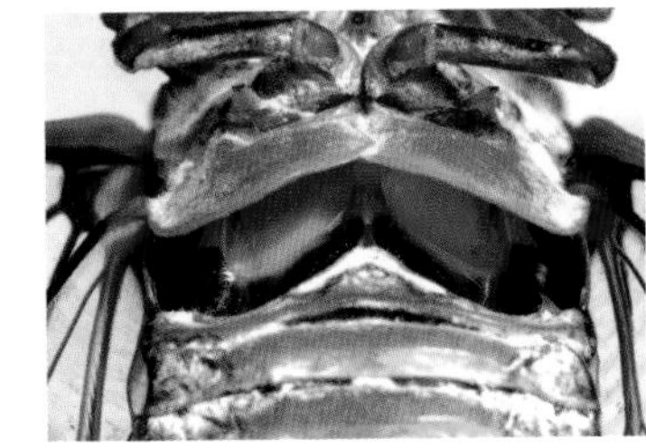

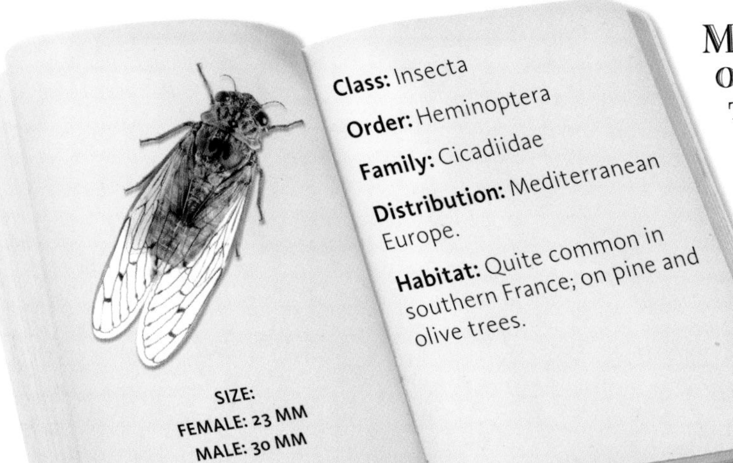

Mistress of camouflage

There are about 15 species of cicadas in France, the Ash Tree Cicada being the most common along with the Plebian Cicada (*Lyristes plebejus*). Contrary to a common idea, cicadas are encountered in almost every region of France, with the exception of Brittany, in southern Normandy and in the North and the Alsace regions. It is well known that cicadas "sing" very little when temperatures dip below 72°F (22°C), and they are always difficult to spot. As soon as you approach a cicada, it will become quiet and remain perfectly still while camouflaging itself on the tree bark (thanks to its transparent wings and cryptic body coloration). As soon as danger has passed, cicadas resume their normal activities, piercing the bark of trees and shrubs with their long rostrum in order to extract the sap.

Origin of the name

The word *Cicada* is the Latin name of this insect (*cigale* in French); it is extracted from the Greek, meaning "rooster," as an analogy to this winged and singing animal. The name of the species, *orni* refers to the ash tree (*Fraxinus ornus*), even though this cicada is more often found on pine and olive trees. The Ash Tree Cicada is also commonly called the Gray Cicada, the Variegated Cicada and the "Cacan" in Provence.

Singing all summer long

From mid-June to mid-August, Southern France echoes with the powerful song of the cicadas, imparting a charming resonance to this region. Only the males sing, and they do so to attract the females or rebuff their rivals. These strident and monotone sounds, sometimes deafening (up to 158 decibels!), are produced by the vibration of specialized structures, the cymbals, situated on each side of the abdomen, in a sound-amplifying cavity. Cicadas are diurnal and sing mostly when the weather is nice. The Ash Tree Cicada sings on tree trunks and large branches, though many other species prefer small branches or new twigs.

Fabre of the Fountain

In his celebrated *Souvenirs Entomomolgiques*, Jean-Henri Fabre got much of his inspiration from the famed author Jean Lafontaine who he admired very much. Referencing Lafontaine's *La Cigale et la Fourmi* (The Cicada and the Ant) he noted that, "the truth rejects as nonsensical invention that which the writer of fables tells us." He had observed that the cicadas did not feed on "bits of flies or of small worms" but that they bit plant materials to extract their sap. Worse yet, he noted that it was the ants that profited shamelessly of this manna: "When it comes to who is the bandit, the ants win out all the time" he concluded.

Ash Tree Cicada larva

Plebian Cicada (Lyristes plebejus)

A long larval lifetime

Following mating, at the end of August or beginning of September, the female inserts her eggs into the stem of a plant with the help of her ovipositor. Come fall, the larvae hatch and fall to the ground where they then burrow. They will live there for many years (from two to six years for *C. orni*) feeding on plant roots. In June, the larvae leave the soil to pupate one last time on a shrub. Adults only live for about a month and half.

Cicada orni

Linnaeus, 1758

Size:

Female: 23 mm

Male: 30 mm

Rural Tiger Beetle larva

Rural Tiger Beetle

SIZE: 12–15 MM

Class: Insecta

Order: Coleoptera

Family: Cicindelidae

Distribution: From Europe to Siberia, and North Africa.

Habitat: Wherever there is barren soil, on roads or clearings.

ORIGIN OF THE NAME

The Latin name *Cicindela* is a doubling of the Latin verb *candere* – to shine – a reference to the often coppery reflections from the surface of their elytra. The species name *campestris*, "of fields," is not quite appropriate for an insect that prefers to frequent sandy roads, sunny embankments and the edges of dried woodlands.

THE TIGER OF ALL INSECTS

While strolling along a sandy roadway and under pleasant, sunny conditions, it is easy to encounter the Rural Tiger Beetle. A bright, shiny green spark will alert you to its presence. Its big globular eyes impart it with an exceptional field of view ... and it takes off as soon as you approach it. When it comes to rest, it seems to suddenly disappear, since its markings display lines that perfectly camouflage it on the soil while hiding its outline.

The Rural Tiger Beetle seeks light (and warmth) to become active. By night, or under cloudy conditions, it hides in its shelter, under vegetation and is completely inactive.

The Cicindelidae are fearsome predators – thus providing the reason why we refer to them as "tiger beetles." They hunt various small insects in flight, or more often by chasing them on the ground. They rapidly chase their hapless victim, which ends up pierced by the tiger beetle's huge serrated mandibles.

SUBTLE HUNTING

Ernst Jünger, a prestigious 20th-century German writer whose literary work is renowned and admired worldwide, was like the Russian poet Nabokov – a confirmed entomologist. In his 1967 book *Subtle Hunting*, he does not hesitate to reveal his passion for insects since, during his long and tumultuous life (he lived to be more than 100 years old), he gathered a collection of coleopterans, notably the Cicindelidae (tiger beetles). It is therefore not surprising that one of the specialists in this genus gave his name to a species: *Cicindela juengeri* (Naviaux, 1992), from the south of India.

BURROWING LARVAE

Springtime is the season for "making love." During mating, the male holds on to the female with its powerful mandibles. The eggs are then laid separately in the ground. The larva of Rural Tiger Beetle develops in sand, in a very deep vertical tunnel (up to 16 inches/40 cm deep) that ends horizontally. Though it does not move around very much, it is difficult to surprise it since it is very sensitive to soil vibrations. At the slightest hint of danger, it digs itself very deep into its tunnel.

At the completion of its development, the larva eats the opening of its tunnel and enlarges the end to pupate. The adult emerges in the following spring. Rural Tiger Beetles have few natural enemies, though humans have contributed significantly to the destruction of their habitat.

A CATERPILLAR'S NIGHTMARE

The Rural Tiger Beetle larva guards the entrance to its vertical tunnel – with its perfectly trimmed outer edges – leaving a hole that is even with the soil's surface. Its head and prothorax form a platform that cleverly hides the entire entrance. Hanging on against the tunnel's walls with two little hooks at the base of its back, very much like a chimney sweep, the beetle awaits the passage of its prey – caterpillars or other soft-bodied worms – that it then grasps quickly with its long, pointy mandibles and carries to the depths of the tunnel where it will be devoured.

Rural Tiger Beetle devouring a fly

Cicindela campestris

Linnaeus, 1758

Size: 12–15 mm

Coccinella septempunctata

Seven-spotted Ladybug

Ladybug nymph

Origin of the name

The name *Coccinella* is a diminutive drawn from the Latin *coccinus*, meaning "scarlet." The species name *septempunctata* ("seven spots") obviously indicates the number of black spots this ladybug carries on its elytra. This also lays to rest the rumor that the number of spots is indicative of the animal's age!

The most popular insect

Particularly easy to observe with its red markings on the surface of green vegetation, the Seven-spotted Ladybug is extremely common and widespread. It can be found nearly all year long in plains and mountains, in forests, fields, parks or gardens – in fact anywhere it can feed on its favorite meal: aphids.

When disturbed, it releases an orange-red, bloodlike substance from its articulations that both smells and tastes bad. This is a clever defensive technique to scare away predators. The ladybug's bright color is also a heads-up to predators that its flavor is unpleasant and that they would be well advised to look somewhere else for their prey.

Organic gardener's ally

Beyond its pleasant and friendly coloration, the ladybug owes much of its popularity to its voracious appetite for aphids. Considered a serious natural ally, it is always welcomed by gardeners, who do not hesitate to voluntarily seed their gardens with these little insects. In fact, the female ladybug deposits small groups of 50 to 200 yellow-orange eggs under the leaves of aphid-infected plants. Following hatching, its larvae will feed for one to two months (depending on the temperature), devouring up to 200 aphids per day up to their last developmental stage. This is followed by the immobile nymph stage, during which the larvae metamorphose into adults.

The adult itself also has a solid appetite. It devours about 100 aphids each day. In summer, when the temperatures rise and aphids become less numerous, ladybugs may migrate en masse. This behavior warranted their German name – *Marienkäfer* ("Marie's beetle") – since the migration occurs often toward the Assumption of the Blessed Virgin Mary (August 15).

At the end of fall, ladybugs are on the lookout for a spot to hibernate, either under leaves, in moss or under rocks, and they may even happen to take up residence in our homes. They will remain in their shelters patiently awaiting the coming of the following spring.

GOD'S LITTLE BEAST ... AND SOME SUPERSTITIONS

The ladybug is supposed to bring luck. In one story, as it goes, a ladybug landed on the neck of a condemned man at the moment of his execution. The King saw this as a sign from God and granted him a pardon. It has also been said that if a ladybug lands on you and immediately takes off again, the next Sunday will have good weather, but if it lands and remains on your finger, rain is in the forecast. If you can count to 22 before this happens, you are destined to be happy! Deposit a ladybug in your hand, make a wish, and make it climb up on your finger extended to the heavens. The insect will then take flight to carry your wish up to God.

The voracious ladybug larva

Dining on an aphid

ACTUAL SIZE

Coccinella septempunctata

Linnaeus, 1758

Size: 5.5–8 mm

Azure Damselfly

Origin of the name

The name "agrion" is derived from the Greek *agrios* (meaning "wild"). The Greek prefix *koneo* (that which lives sociably), added to the genus name, refers to this insect's behavior. The species name *puella* means "young woman" and follows the tradition of the first names given to the damselflies.

A very pretty miss

The Azure Damselfly is one of the most common damselflies (all of the order Zygoptera). It is mostly found by the shores of stagnant or still waters, often in great numbers. The abdomen is sky blue with black rings, and generally has a U marking closest to the wings (second abdominal segment). The female, identified by her paler abdomen, only lives on the water when laying her eggs or mating. On the other hand, the male, even though he isn't quite as territorial as the Damselfly (p. 58), always remains close to the water and never flies.

Morning nymph stage

The larva of the Azure Damselfly lives in ponds, lakes and marshes at elevations of up to 6,500 feet (2,000 m). It stays within aquatic vegetation or along the bottom and moves slowly, occasionally chasing a passing prey. Other related species have less strict ecological needs, as in the case of the magnificient Large Red Damselfly (*Pyrrhosoma nymphula*, above) that develops just as well in still waters as it does in small, clear and rapid flowing streams.

It takes approximately from six months to two years for the larva of the Azure Damselfly to complete its development. In the spring, usually in the morning, it leaves the water to latch on to an embankment plant. If temperatures permit, its skin will split along the thorax to permit the adult to emerge. If conditions are not ideal, it will return to the water to wait for a more propitious time. Once the deployed wings have dried (a process taking up to about two hours), the damselfly is free to fly away.

Acrobatic coupling

Reproduction occurs from May to August. During mating, the male grasps the female behind her head with its anal appendages, while she bends her stomach to bring her genital orifice into contact with the male's copulatory organ. The male remains this way – posed as a "sentinel," hooked onto the female until she lays her eggs (image below). Egg-laying takes place in emerging aquatic plants only in good weather. Larvae will hatch from three to five weeks later.

DEMONS AND WONDERS

In Japan, the damselflies, symbols of lightness and elegance, are harbingers of happiness. Emperor Jimmu Tennõ, the mythological founder of Japan, is thought to have baptized his country Akitsu Shima, "damselfly island," because the region of the archipelago that he could view was evocative of two mating damselflies. For Native Americans, the damselfly symbolizes the deception of the senses. Its brilliant wings make us aware that our world is but a façade. In the West, ancient beliefs considered the damselflies as manifestations of evil spirits, because they inhabited aquatic environments thought to be haunted. A Walloon legend even maintains that when a damselfly hits someone on the forehead, death follows within the year!

Male (blue) and female mating

Coenagrion puella

(Linnaeus, 1758)

Size: 34 mm

Wingspan: 50 mm

Brown Bug

Class: Insecta

Order: Heminoptera

Family: Coreidae

Distribution: Europe, Near East.

Habitat: Everywhere they can find their favorite plants, the docks and sorrels of the Rumex genus.

SIZE: 10-14 MM

ORIGIN OF THE NAME

The Brown Bug, Dock Bug or Squash Bug, also known as the Marginated Corea, owes its species name *marginatus* to the fact that the abdomen flows out laterally from the wings and forms a sharp border or margin. The etymology of *Coreus* certainly refers to the Greek word for bug, *Koris*, though it could also be an allusion to its shape and color (*corium* means "copper" in Latin).

A TASTE FOR SAP

Bugs from the Coreidae family (more than 2,000 species in the world) are for the most part tropical, although several species are found in temperate zones. The Brown or Squash Bug can be observed from May to October in humid areas, at the edges of forests, clearings and the ditches where their favorite plants, sorrel and impatiens (genus *Rumex*) grow. The juveniles feed on the sap and sugars in leaves, while the adults feed on the seeds. It is also very common to encounter these insects on branches full of blackberries and on rhubarb.

CHARACTERISTIC SILHOUETTE

The color of this large bug varies from a light yellow-brown to a deep red-brown, with a grey-brown in between. Generally, an individual darkens progressively during the course of the year. Its very large pronotum gave it the old genus name of Syromastes (from the Greek, meaning enlarged thorax or chest). The posterior region of the abdomen, which is very dilated, extends on the sides of the elytra, giving this beetle its characteristic shape. A closer look will reveal that the head bears two small "horns" poking up between the bases of the antennae.

WELL HIDDEN EGGS

Reproduction occurs in May and June. The male and the female communicate via pheromones emitted from the odiferous glands positioned on the side of the thorax; these same substances also produce the nauseating substances that give these bugs their bad reputation. Mating occurs back to back, as with the Firebug (p. 208). The fertilized female lays her eggs apart from all other insects on the host plant. The eggs can easily be confused with the tiny fruit of the *Rumex*, which is the buckwheat family of plants, such as docks and sorrels (see photos above). The juvenile insects are wingless and their antennae appear disproportional (see photo below). Their odiferous glands are located underneath the abdomen, where they form two dark, somber markings. The young develop in the summer and fall, then hibernate. They will attain maturity in May of the following year. They will progressively acquire the adult's morphology after several successful molts.

AN OVERLOOKED HELPER

The Brown Bug causes only minor damage in gardens, since the plants they truly prefer are not cultivated. It goes without saying that its lingering odor sometimes "perfumes" the fruit and detracts from the pleasure of savoring them! However, this bug could easily become a valuable asset to farmers in their effort to control the proliferation of a noxious weed the Broad-leaved Dock (*Rumex obtusifolia*) a particularly invasive species. Studies are underway to explore this possibility.

Eggs well camouflaged amid sorrel seeds

Larva of the Brown Bug

Coreus marginatus

(Linnaeus, 1758)

Size: 10–14 mm

ACTUAL SIZE

Sycamore Lace Bug

Class: Insecta

Order: Heminoptera

Family: Tingidae

Distribution: Originated in North America; currently widespread in central and western Europe, as well as Japan.

Habitat: On sycamores.

SIZE: 3–4 MM

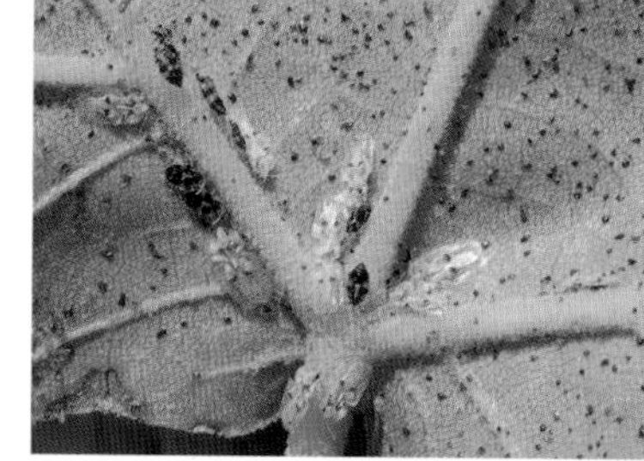

Larva and adults

Small but ferocious

The adults hibernate under small bark plaques (the rhytidomes) that are very characteristic of sycamore tree trunks. Come springtime, they awake and migrate toward the tree's leaves. They settle on the underside of the leaves, close to the central vein, and begin to actively pump out the sap with their biting rostrum. About ten days after that, they mate. The females lay between 200 and 300 eggs. About 20 days later, the larvae emerge, resembling adults, though without wings. Following four molts over some 30 days, they then become beautiful adult bugs, festooned with lace. During the following season, two generations (three in warmer regions) follow in succession, leaving adults, eggs, larvae, molts and excrement on the underside of the sycamore's leaves.

Leaf of stricken sycamore

Origin of the name

These Lace Bugs are superb insects of the Tingidae family, so small that it is quite difficult to appreciate their beauty. This species derives its genus name from the swollen shape of its pronotum that covers the head very much like a little helmet (*koruthos*, in Greek). The species name, *ciliate*, "ciliated," refers undoubtedly to the little spiny structures at the edging of the anterior section of the pronotum.

Fearsome lace bugs

It would appear that we should fear the Lace Bug found on the Swiss oak tree, since the related species, (*Stephanitis piri*), which is found on pear trees, has long been the most well known and feared of all Lace Bugs. However, it seems that the Sycamore Lace Bug threatens to soon outstrip its predecessors.

Originating in North America, the Sycamore Lace Bug was accidentally imported to Europe. First noted in Italy in 1964, it was later identified in the Antibes region of France in 1975. This species has spread incessantly since then throughout Europe, both north and east. In 2001, it was also reported in Japan.

SOS: endangered sycamores!

This lovely little insect has become quite a headache for green space developers who have planted long rows of stately sycamores. The first sign of infestation is the appearance of small discolored spots around the veins on the upper surfaces of leaves, which cause a lot of damage. This, coupled with honeydew excrement on the already weakened trees, promotes the development of the fungi responsible for alarming plant illnesses, such as sycamore canker.

The Lace Bugs are also responsible for other nuisances: they land on people basking on terraces in the shade of sycamore trees, and they stick to joggers running along canals near infested woodlands.

DIVERSIFY OR DIE

It is very difficult to stamp out the Lace Bug infestations either with pesticides or with biological methods. Though these devourers have some natural predators like the mealy bugs and the Chrysope that attack their larvae these come too late to really contain the population explosions. The best solution consists in diversifying the species of trees we plant so as to diminish the risks of insect proliferation and overpopulation.

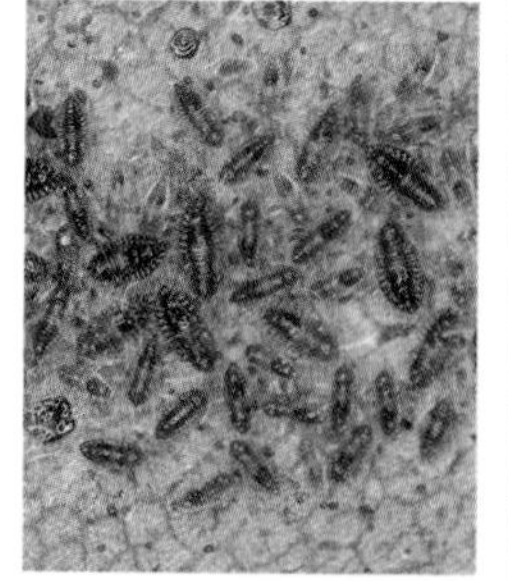

Young larvae

Corythucha ciliata

(Say, 1832)

Size: 3–4 mm

ACTUAL SIZE

Cabbage White Butterfly Parasite

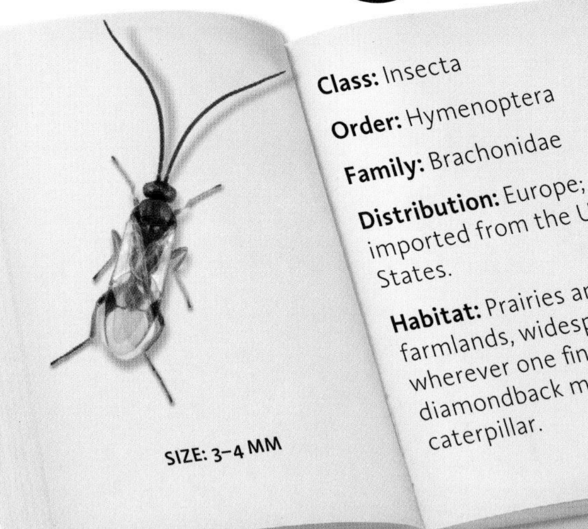

ORIGIN OF THE NAME

Apanteles is the old scientific genus name of this little hymenopteran. It literally means "achievement without total." It is also often called the Microgaster ("small stomach" in Greek: a reference to the reduced size of the adult stomach). The name was finally settled in 1891, as the genus *Cotesia*. Finally, the species name *glomerata* means "gathered" in Latin, which refers to its gregarious method of development.

A SMALL "ALIEN" LIVES CLOSE BY

The Cabbage White Butterfly Parasite is a parasite of the cabbageworm caterpillar (p. 202), or rather a "parasitoid:" a parasite that systematically destroys its host. Using its long ovipositor as a syringe, the female lays about 20 eggs in the young caterpillars of the cabbageworm butterfly. The larvae hatch and proceed to feed on the blood (hemolymph) and the fatty bodies of their victims. On first feeding, they spare the vital organs. The caterpillars thus remain alive and continue their development, despite the presence of numerous clandestine passengers that are devouring them from the inside out. Things go very badly after completion of the fifth larval stage, at which point the parasitoids have completed their own development. They leave their hosts by piercing the tegument (skin) and then proceed to weave small yellow cocoons hooked into the cabbage leaves, close to the cadaver of the caterpillars.

Laying eggs inside a caterpillar

THE SPRAYER IS SPRAYED

Seemingly sheltered in their golden cocoons, the transformation of the Cabbage White Butterfly larvae into adults lasts only for a few days during summer. However, this does not take into account the "hyperparasitoids" such as Lysibia nana: the females of this little insect seek out these golden cocoons and lay their eggs within them, thereby assuring that their own larvae will survive at the expense of those of the Cabbage White Butterfly!

CALL OF THE CABBAGES

Cabbages are not totally vulnerable to attacks by herbivorous insects. When attacked by the cabbageworm, they emit chemical substances that attract the enemies of their aggressors, such as the Cabbage White Butterfly Parasite. These chemical messages are synthesized by the plant in response to certain compounds found in the caterpillar's saliva. These signals begin in the plant's caterpillar-damaged areas, but they are given off by the entire plant, thereby increasing the range and probability that the signal will detected by the parasitoids.

THE PARASITOIDS OUR FRIENDS

These small insects play a critica role in controlling a species harmful to crops. In cultivated fields, *Cotesia glomerata* can parasitize up to 95 percent of the cabbageworm caterpillar! These exceptional results guarantee the rapid eradication of cabbageworms without having recourse to pesticides. This has been known for a long time: the Cabbage White Butterfly Parasite has been used to combat the cabbage-ravaging caterpillars since 1883. Another species, *Cotesia rubecula*, plays the role of a complimentary "agricultural assistant," since its larvae kill cabbageworms at an earlier stage than those of the *C. glomerata*.

Emergence (to the right) and cocoon (below)

Cotesia glomerata

(Linnaeus, 1758)

Size: 3–4 mm

Ctenolepisma lineata

The four-lined silverfish

Four-lined Silverfish

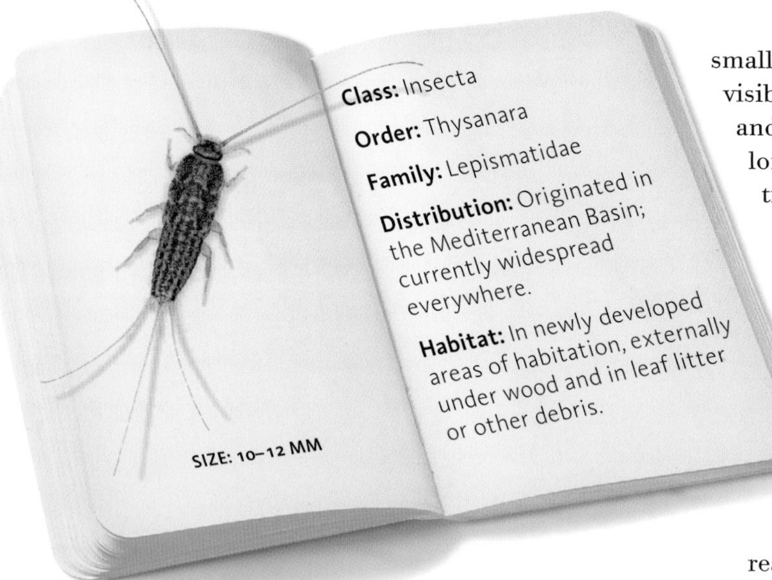

Origin of the name

The name of the genus, *Ctenolepisma*, means literally "comb-carrying silverfish," with the word itself derived from the Latin *lepisma*, "scale," in reference to the scales that cover the body of this insect. The species name *lineate* refers to the four longitudinal black striations that line the insect's back. This insect is also known as Bristletails.

Small table companion

Very agile, and essentially nocturnal, the Four-lined Silverfish dwell in habitations. They are often encountered on old papers, on which they feed. People often surprise them on their balconies while lifting an object from the floor. Silverfish also live in natural environments, notably under tree bark, in more southern regions. An interesting aspect of their behavior is that they can go without water or food for several months.

Poorly known insects

Bristletails are primitive insects without wings, with a spindle-shaped body covered in scales. This morphology has given them the common name of "silverfish." Their masticating mouthparts allow them to consume a variety of foods, most often in the form of small detritus material. The thorax has three visible segments that bear three pairs of legs, and the elongated abdomen that ends in two long, lateral, thin appendages and one central epiprocte (supra-anal shield). The old name of this insect order (Thysanara), referring to *Thysano*, "fringe" and *ura*, "tail," was given because of the sensory covering on their posterior appendages. Bristletails grow slowly and have a long life expectancy (up to four years). Females lay their eggs directly on the ground, in the litter or in minuscule fissures. Incubation times (ranging from one to nine weeks) depend largely on ambient temperatures. The juveniles resemble adults, although in a smaller form. They reach sexual maturity after a few months. One phenomena that is rarely seen in insects is the continuous molting of the adults; an individual can molt up to 70 times in its lifetime! Some species live in anthills or termite mounds, where they behave mostly as housemates.

Before the appearance of wings

Not long ago, the bristletails were joined to other archaic wingless insects (the Machilidae), among them the "Apterygotae" (without wings). The absence of wings here is not an adaptation but more a primitive characteristic. They did not lose their wings secondarily like ants: they never had any! It was therefore an artificial grouping and today bristletails are placed in the order Thysanara and the Machilidae are placed in the separate order of the Archeognathae.

Silverfish

Other than the four-lined silverfish, two other Bristletails dwell in habitations: the common Silverfish (*Lepisma saccharina*) and the domestic Firebrat (*Thermobia domestica*). These two were originally tropical, but are now widespread more or less throughout the world, the former in habitations and the latter in warm buildings like bakeries. The silverfish has a body covered in silver scales hence its name. A nocturnal insect, it is quite content to feed off old paperwork. This insect can cause quite a bit of damage in bookstores and libraries, since is favors the flour that makes up most of the glue in old book bindings.

Sheet of paper nibbled by Bristletails

Marauding on wood

ACTUAL SIZE

Ctenolepisma lineata

(Fabricius, 1775)

Size: 10–12 mm

Curculio nucum

Hazelnut Weevil

A weevil's sting

Origin of the name

The word *balanin* (French for weevil) is from the Greek *balanos*, "acorn," and refers to the larva's habitat and applies to weevils (*Curculio* in Latin) equipped with a long, thin, bent "nose." The name of the species, *nucum* ("from nuts"), refers to its habitat.

Expert in foraging

Who can forget the experience of biting into a freshly picked hazelnut and discovering a wiggly larva in it, working hard to transform this delicious treat into bitter dust that we spit out with disgust?

It is at the beginning of fall, when hazelnuts are still green that the female weevil lays her eggs. The odd appendage, shaped like a trumpet, which emerges from between two big, dark eyes gives the weevil its comic appearance. This rostrum, ending with two tiny, sharp mandibles, turns out to be a very effective foraging tool. Still, it takes several hours of digging before it arrives at its target site – the core of a hazelnut (Jean-Henri Fabre gives this a delicious description). The female then inserts her ovipositor, a long, rigid tube normally hidden inside of her, into the hazelnut. She slips her egg in and it travels to the depths of the tiny tunnel. The irritation caused by the sting causes the nut to produce new plant tissue, which will form a scar over the sting's opening.

A patient larva

After an incubation lasting from four to nine days, a miniscule "worm" begins to gnaw at the interior of its capsule. It grows into a large larva and at the end of the month, it extirpates itself from its cell by the means of biting little round holes with its mandibles. Meanwhile, the hazelnut meets its demise prematurely and falls to the ground. The larva buries itself to a depth of four to six inches (10 to 15 cm) and builds a shelter in which it will hibernate during two or three winters. The adults will then emerge at the end of summer and the females will lay their eggs in the fall. The development cycle of this insect necessitates a period of two to four years.

To each his own fruit

The larvae only develop in hazelnuts, though the adults are less choosy about their habitats. They do appreciate young hazelnuts (although the males prefer to devour the leaves), but they also love to dig their rostrums into juicy fruit, notably peaches and pears. Not having to dig quite as deep as the females, the males have a much shorter rostrum. In some species, notably the Elephant Weevil (*C. elephas*), which attacks chestnuts, the female's rostrum is longer than the rest of her body.

A BIG FAMILY

Our weevil belongs to the huge Order of Coleopterans (350,000 species!). It is estimated that the family of weevils (Circulionidae, in scientific terms) has 60,000 species – more than all species of fish, amphibians, reptiles, birds and mammals together!

The strange "look" of the hazelnut weevil

Curculio nucum

Linnaeus, 1758

Size: 6–9 mm

ACTUAL SIZE

Birch Leaf Roller Beetle

Class: Insecta
Order: Coleoptera
Family: Rhynchitidae
Distribution: Temperate and hot Eurasia; North Africa.
Habitat: On numerous forest trees: birch, beech, alder, hazel tree …

SIZE: 2.5–4 MM (INCLUDING THE ROSTRUM)

Rolled leaves in the shape of cigars

Origin of the name

The Latin name *rhynchites* originated from Greek, meaning "beak:" an allusion to the large rostrum of these insects. *Deporus* is currently a genus name separated from the Rhynchitae. The species name refers to the birch tree (*betula* in Latin). This weevil and other similar species are currently called "cigar-makers" because they roll the leaves in the form of a cigar around their eggs.

On the quest for a good birch tree

The Birch Leaf Roller Beetle belongs to an immense family of weevils, as does the Hazelnut Beetle (p. 100). The characteristic long rostrum is evidence of this linkage. The male distinguishes himself from the female by swollen posterior femurs. Forest, parks, gardens, wood-like swamps and their surroundings constitute its domain, from springtime to the beginning of summer. As its name indicates, this beetle is linked to the birch tree, although it can also develop on alders, the hazel tree, the beech and many others.

How to roll a cigar

The female of this species lays an egg on the undersurface of a birch (or other tree) leaf and transversely severs the limb, leaving only the median vein. She then rolls up one end into a thin cone shape. This task, colossal for such a small insect, takes from 30 to 60 minutes. First suspended from the tree, the cone eventually falls to the ground. On the ground, the larva devours the leaf's parenchyma tissue. Then, by the end of summer, it buries itself in the ground, where it will metamorphose in the shelter of a cavity. The adult version emerges two weeks later, although it remains in this subterranean shelter to hibernate. It does not emerge from the ground until springtime, once the new young leaves have formed on the tree.

Some beetles take short cuts

In Nordic countries, notably in peat bogs, one can encounter a closely-related species, *Deporus mannerheimii*, which is smaller and sports a green or metallic-blue color. This beetle also develops on the birch tree, but also in the willow and the hazelnut tree. One notable exception is that it does not build cigars – the larva chews on the leaf's limb, from which the petiole has already been devoured by the female. The leaf falls quickly and the larva continues its development in a pile of wet leaves on the ground.

The larva in its cigar

ANOTHER ENTHUSIASTIC CIGAR-MAKER

The cigar-maker *Byctiscus betulae* (left), despite its species name ("of the birch"), lives at the expense of many different forest trees and freely attacks vine leaves. The male is armed with two spines on its pronotum. Larger (2 to 2.75 inches/5 to 7 cm) than that of the Birch Leaf Roller Beetle, the female can undertake a colossal task: she "rolls" 20 to 30 cigars (with one egg laid in each cigar), at the rate of two per day! The technique of cigar fabrication varies from one species to another, and the form of the cigar is a good indicator as to what species made it.

Deporaus betulae

(Linnaeus, 1758)

Size: 2.5–4 mm

(including the rostrum)

Wild Rose Gall Wasp

The "hairy" g

Origin of the Name

The Wild Rose Gall is also called the Bedeguar Gall and the Moss Gall. The French name *Cynipe* (which comes from the Greek *Kunos*, "dog," and *ips*, "worm") justly refers to this species that develops on the wild rose (*Rosa canina*), once called the Dog Rose. It was once thought that one could utilize its roots (or galls) to treat rabies in cases of dog bites. The name of the genus *Diplolepis* ("double scaling" in Greek) goes back to the two laths into which the ovipositor is enrolled. The very beautiful "hairy" gall made by the insect is named "St-Peter's Beard," or *Bédégar* (from the Arab-Persian *Bâdaward*, meaning "rose (carried by) the wind"). Its shaggy-haired appearance also gives rise to another common name, "Robin's pincushion," in reference to the red-colored feathers of that bird.

But Where Are the Males?

Each gall is host to tens of larvae, almost all of the female gender: we only count one male for several hundreds of females! This extreme scarcity of males is linked to the capacity of females to reproduce by parthenogenesis, which means that they don't need to be fertilized to lay viable eggs. Usually leaving the galls in springtime, the females go hunting for new tree buds, into which they will insert their eggs with the help of their ovipositor. The larvae that hatch will then enter the plant's tissues, resulting in the formation of a tree gall. The latter will reach its maximum size (from .75 to 2 inches/2 to 5 cm, depending on the number of larvae) within four to eight weeks. Under its wild hair, the gall hides several loges, each containing one larva, with the walls of the loges serving as nutrition. The larvae spend the winter in these loges and molt into nymphs by springtime before emerging. The adults escape from the loges by digging tiny tunnels in the plant tissue with their mandibles. And the cycle resumes one more time...

Popular and Surprising Practices

In days gone by, we attributed all sorts of healing properties to these galls. They were once used to cure digestive ailments and urinary troubles. They were put under the tongues of colicky babies or under their pillows as sleep aids. They were also used as talismans to protect a subject from toothaches or hemorrhoids! The dried, pulverized galls have, on occasion, been used as substitutes for pipe tobacco. In Italian churches, they were even used as holy water!

All Kinds of Squatters

The galls play hosts not only to the larvae of these wasps but also to a micro-community dependant on these galls and their occupants. As such, we have counted more than ten other species of Hymenopterans. Some, said to be commensal, who unable to induce the formation of galls themselves, feed on formed plant tissues without causing harm to their constructing proprietors. Other are predators and devour the larvae therein. Still others are parasitoids: their own larvae develop and grow on the larvae of the wasp or the commensal larvae. And, once the gall is empty of its diverse occupants, it is not unusual to notice that spiders then install themselves in the newly-vacated gall!

Larvae in a gall

Egg-laying (left) and Parasitic Wasp (right) (Orthopelma mediator)

ACTUAL SIZE
N. Diplolepis rosae
(Linnaeus, 1758)
Size:
3.8–4.9 mm

Dolomedes fimbriatus

Raft Spider

The Raft Spid
beside its last m

Young Raft Spiders

Origin of the name

The genus name *Dolomedes,* created by Latreille, means "cunning" in Greek. The name of the species, *fimbriatus*, means "fringed," a reference to the clear margins of the cephalothorax and abdomen.

An attentive mom

In April, the Raft Spiders mate. The male shakes his front legs one after the other towards the female, to attract her attention – which he hopes will be benevolent. This rather long parade is followed by a very brief mating, during which the male uses only his "boxing gloves" (masticating limbs transformed temporarily into mating organs). In June, the female lays several hundreds of eggs, spread between two to three greenish cocoons. During incubation, she transports her awkward progeny in her chelicerae and under her sternum. When the young appear ready to hatch in the middle of summer, she suspends the cocoons in low-lying vegetation and weaves a very fine nursery web. She stands guard over this web, and the young remain in this "nursery" until they have gone through two successive molts. The young then disperse by climbing in the foliage of bushes. When they have become adults, they remain near still waters. They hibernate at the beginning of fall.

A family of hunting spiders

When threatened, the Raft Spider immediately dives under water and hides in aquatic vegetation. It can remain immersed and still for half an hour, breathing with the help of air bubbles captured in its hairs.

The *Pisauridae* family has three notable species: two raft spiders (the second being *Dolomedes plantarius*, an endangered species) and the Admirable (*Pisaura mirabilis*), which can be found in herbaceous areas. The latter is distinguished by its odd behavior during mating. The male, in an effort not to be devoured by the female, attempts to calm her down by offering her a fly wrapped in fine silk. Theoretically, he can then mate while she is preoccupied with her "gift."

The Raft Spiders, and more generally the *Pisauridae*, are armed with powerful chelicerae (below). Better not try to capture them, since their bite can be quite painful.

IT WALKS ON WATER

The female Raft Spider stays on vegetation bordering a pond or a lake, with her forelegs on the water's surface. She can then detect any passing preys, even making water motions to attract them. As soon as a prey appears – damselfly, tadpole or even a young fish – she launches herself on the water and runs along its surface. She does not hesitate to dive in, while grasping at the vegetation, to capture her aquatic preys, which she heaves out of the water and devours.

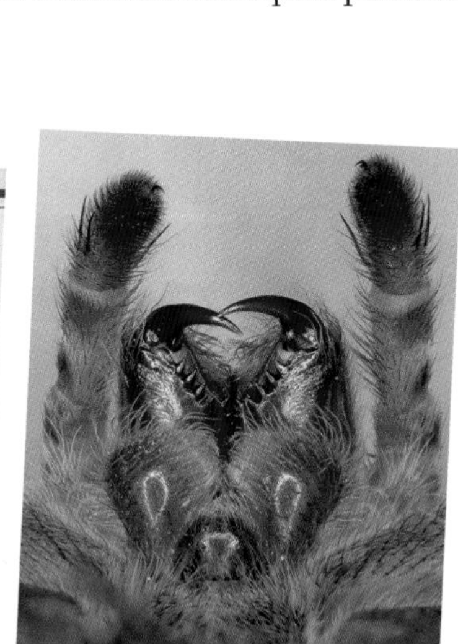

The impressive chelicerates of the Raft Spider

Dolomedes fimbriatus

(Clerck, 1758)

Size:

Female: 16–22 mm

Male: 10–13 mm

Drosophila "Tuba" eggs

Common Fruit Fly

Origin of the name

In Greek, *Drosophila*, means "which likes the morning dew" (*la rosée* in French). As poetic as this may sound, the reality is that the Fruit Fly is mostly attracted to alcohol vapors (hence its alternate name "Vinegar Fly.") It would therefore be more accurate to mention that it likes the "rosé" (as in the wine!). The species name, *melanogaster*, means simply "black stomach." In the male of the species (and there are more than 3,000 different species worldwide!), the extremity of the abdomen is indeed black.

You can't catch flies with vinegar

The Drosophila makes a lie of this famous saying, since they come in droves at the slightest hint of the smell of wine (be it the cheapest or the costliest – it matters not to the Fruit Fly). Not that the adults appreciate the alcohol (they don't consume any), but their larvae develop in decomposing fruit. The fermenting sugar produces the alcohol, a volatile substance that transmits an alert over long distances, marking a favorable egg-laying site. It's not surprising therefore, that the Common Fruit Flies are led astray by the large container of sangria that one brings outdoors on a warm summer night!

It doesn't drink but it shakes

In nature, the Fruit Flies usually gather around a rotten fruit that has fallen to the ground. They feed on the rot that grows from the surface of the fallen fruit. On this unusual meeting place, the male takes advantage of the situation to woo the female, by frenetically shaking his wings in a codified rhythm particular to each species. The female lays her eggs in small crevices. She can produce several hundred over the course of her lifetime. The Drosophila egg is instantly recognizable: two respiratory orifices emerge over the middle of the more or less viscous mass in which they were laid. If conditions are ideal (77°F/25°C), the miniscule maggots hatch a mere 24 hours later. They go about digging methodically around them with hooks that extend from their mouths. Their saliva takes care of a part of the digestion from the outside, and the resultant pulp is swallowed as soon as it is produced.

An accelerated life cycle

Since the medium in which the Drosophila develops is quite ephemeral, becoming an adult is a run against the clock. During the early stages of embryonic development, the cells undergo division every ten minutes, a faster rate even than bacteria! Drosophila goes through its life cycle of three larval stages in only ten days, if the temperature is ideal. The last stage is a rigid chrysalis (the pupa), from which the adults emerge.

THE ELDORADO OF GENETICISTS

Since it reproduces at a lighting speed, Drosophila is a choice "guinea pig" for geneticists. A large part of what we know about mutations has been discovered in Drosophila, studied for more than a century in laboratories around the world. Thanks to this small insect, we have recently identified the genes that control the placement of body parts. If these genes are displaced on their chromosome, a leg will grow where an antennae should or vice versa!

Drosophila pupa (nymph)

The Drosophila maggot

Drosophila melanogaster

Meigen, 1830

Size: 2–4 mm

ACTUAL SIZE

Great Water Beetle

Great Water Beetle larva attacking a tadpole

FEROCIOUS AT ANY AGE

The Great Water Beetle enjoys stagnant water rich in vegetation. This large carnivore attacks mostly tadpoles, newts, aquatic larvae or mollusks, both in its larval and adult stage. However, it also participates in the cleaning of fish cadavers, batrachians (frogs and toads) and mammals that pollute the area. Though the diet of larvae does not differ much from that of the adults, their methods of eating differ considerably. The adult holds its prey between its front legs and tears it apart with its mandibles. The larva has a closed mouth. It perforates its prey with long, serrated mandibles, injects its digestive juice and then proceeds to suck up the pre-digested tissues, leaving behind only an empty shell. Be wary of the mandibles of this insect's larva, as they can really pinch fingers!

ORIGIN OF THE NAME

The name of the Dyticus means "diver" in Greek, originating from this insect's ability to dive to the bottom of ponds and lakes without endangering themselves. Its species name, *marginalis*, refers to the yellow margin that lines the edge of the pronotum and the elytra.

AN UNPARALLELED DIVER

The Great Water Beetle is one of the largest aquatic coleopterans. Since it does not have any gills for respiration, it must often surface to breathe. By placing itself just below the water's surface, with its body inclined at a 45 degree angle to the bottom and its hindlegs widely spread out, it traps air under its elytra. This allows it to breathe while underwater, with the help of respiratory openings located under its membranous wings. Being an excellent swimmer, the adult propels its elongated body using its hindlegs (edged with a hairy fringe) as oars.

FROM THE EGG TO THE ADULT

After mating, usually in springtime, the female buries a small tube, the ovipositor, in an aquatic plant to lay her eggs. The eggs hatch a dozen days later, unless they are laid in the fall. In that case, the eggs will wait till the following spring to hatch. Larvae begin hunting very quickly after hatching. They molt twice until they as big as 2 inches (5 cm). This process that can take as little as twenty days. Then they are ready to become nymphs, a phase that varies according to the season. First the larva leaves the water and sets up house in a small loge on the embankment. Once it reaches the adult stage however, this insect will return to its aquatic environment.

A female Great Water Beetle

TALE OF THE SUCTION CUPS

The forelegs of males are equipped with powerful suction cups (pads) that serve to hold onto the female's elytra during mating (opposite). The female's legs lack suction cups: a handy criterion when attempting to distinguish females from males. The presence of striations on the elytra of these ladies is another identifying characteristic, although less reliable, since in rare instances it is also present in males.

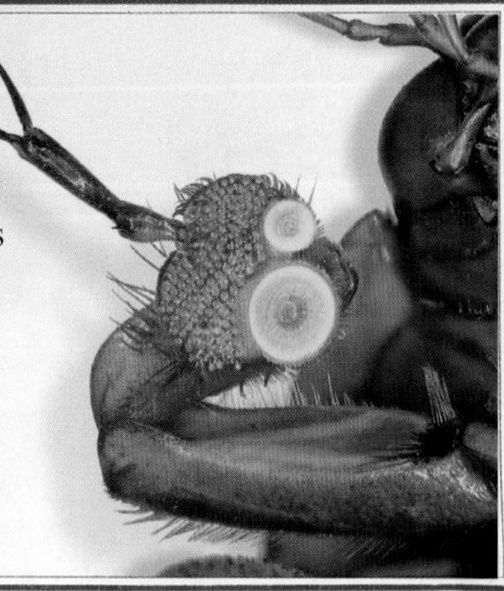

Dytiscus marginalis

Linnaeus, 1758

Size:

25–35 mm

Hairy Rove Beetle

Staphylin larva (Ontholestes)

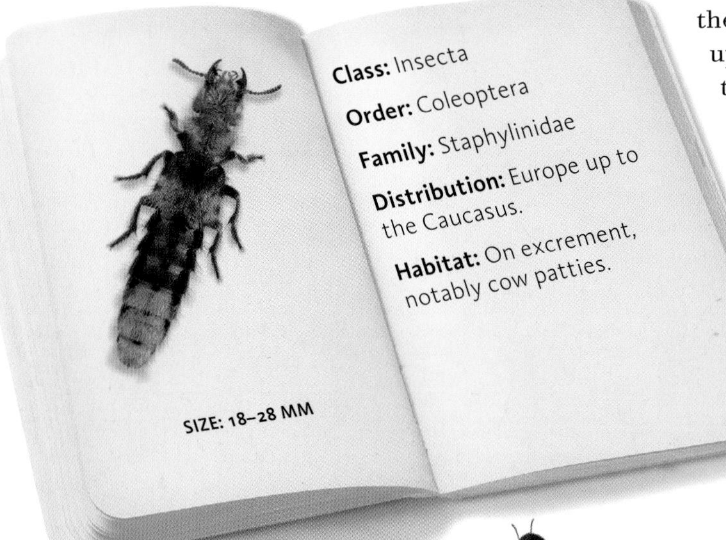

Origin of the name

This beetle is also known as the Staphylin, which is derived from the genus name *Staphylinus*, coined by Linnaeus, who took the Greek name of a non-identified insect (*staphylinos*). Emus is derived from the Greek name of a European freshwater turtle (*Emys*). Today, it is difficult to say just what link Linnaeus saw with our insect. The species name, *hirtus*, on the other hand, does not suffer from any ambiguity. It means "hairy" and refers to the bicolor "hairstyle" of this insect, whose vigorous flight pattern reminds us of a Bumblebee.

Wings in a glove box

The Staphylins consist of no less than 30,000 species worldwide. These insects are easily distinguished from other coleopterans: they have a flattened body, a large head and elytra so short that they only cover a small portion of their elongated abdomens. Since their wings are usually well developed, they have their own unique way of folding them under their elytra: they bend their abdomens upward so as to delicately push their wings under the miniscule "cases" (see photo). Most of the Staphylins also raise their abdomens vertically when threatened, like a scorpion's tail, while emitting a disagreeable odor.

Creophilus maxillosus folding its wings

A feeder on feeders

The Hairy Rove Beetle is hard to catch by surprise, since it flies quickly in the bright sunshine to land on fresh manure piles and escapes at the slightest threat. This carnivore feeds on insects that lay their eggs in and feed on manure, particularly on dipteran larvae. The larvae of Hairy Rove Beetles behave in the same manner as their adult counterparts. The size of adults varies considerably, depending on the region where they live and the availability of food. These beetles have very large mandibles (see *Creophilus maxillosus*, left) and while they prefer manure piles, they have also been observed to feed on carrion.

A poorly understood small world

Because of their distinct behavioral patterns and the large number of very small species involved – some are among the smallest coleopterans in the world (0.5 mm in length) – the Staphylinidae remain poorly understood. Their patterns of behavior are nevertheless quite diversified. In addition to the predators, such as the Hairy Rove Beetle, we also encounter species that feed on mushrooms, cadavers and mold spores. Others inhabit muddy riverbanks or waterfalls, while others are found in mammalian habitats or bird nests. The most unusual Staphylinidae live in ant hills, and even though they feed on ant larvae, the ants do not appear to bother them!

A DISAPPEARING INSECT

The Hairy Rove Beetle is quite rare and appears to be endangered. One assumes that different products fed to beef and livestock – worm powder and antibiotics – have something to do with this, since their byproducts are found in the feces. The same phenomenon is evident in all copraphagous insects. Cow patties and dung piles will accumulate in fields and meadows if these insects can no longer develop therein.

Portrait of the Hairy Rove Beetle

ACTUAL SIZE

Emus hirtus

(Linnaeus, 1758)

Size:

18–28 mm

Episyrphus balteatus

Hoverfly

Hoverfly nymph (pupa)

Origin of the name

Also know as the Marmalade Fly, which alludes to the ornamentation on the abdomen, as does the species name, which originates from the Latin *balteus*, "centurion." The genus name *Episyrphus* refers to its lineage with the genus *Syrphus* (from the Greek *epi*, meaning "on," "over," "plus"). The word *"syrphe"* itself originates from the Greek *surphos*, "fly."

False wasps, true imitators

Each year, the Hoverfly appears in large numbers on wildflowers and in gardens. This little graceful fly with a buzzing flight pattern is easily confused with the solitary wasp. Many Hoverflies mimic the hymenopterans they share flowers with. The Hoverflies mimic the solitary wasps both in terms of size and in their abdominal pattern of black stripes against a lighter background. It is likely that this mimicry dissuades potential predators and fools many hymenopterans, who do not attack them. The imitation is so realistic in some hoverflies that one hesitates to grab them, even though we know that they are only flies! Upon close examination, however, it becomes apparent that these insects do not have two pairs of wings like hymenopterans, but do have the little characteristic balancing structures of dipterans.

Anti-aphid armament

The Hoverfly larva (and that of many other syrphes) is a predator of aphids as implacable as the ladybug. Its controlled use in vineyards could limit the use of pesticides. Stimulated by the volatile substances emitted by plants attacked by aphids, the female hoverfly approaches and blatantly lays her white, elongated eggs. After hatching, the larva, which is initially blind, sets about assailing aphids by moving its head from right to left. One by one it empties the colony of its resident aphids. A month later, at the end of its growth phase (after having consumed up to 1,200 aphids!) the larva metamorphoses into a small pupa in the shape of a water droplet. Unlike the ladybug, the adult Hoverfly does not consume aphids, choosing instead to feed on the nectar of flowers.

An established migrator

Most Hoverflies hibernate in their larval state and metamorphose in springtime. It also happens that fertilized females hibernate and re-appear at the onset of warm days. Many migrate to the south in winter, and in summer the inverse occurs.

STATIONARY FLIGHT

Imagine a helicopter that goes from a stationary flight to a speed of 3,000 mph (5,000 km/h) in one second: at top speed it would reach Mach 14 (14 times the speed of sound). These are the extraordinary feats, proportionally of course, that the Hoverflies are capable of. While hovering, their wings beat so quickly (300 times per second!) that they can become invisible or appear to be levitating. Then, in the blink of an eye, they propel themselves distances of several feet (meters).

Larva devouring aphids

The females' eyes are separate, while those of the male touch each another.

ACTUAL SIZE

Episyrphus balteatus

(DeGeer, 1776)

Size: 8–12 mm

Parisian Antlion

The Antlion larva

Class: Insecta

Order: Neuroptera

Family: Myrmeleontidae

Distribution: From Europe to the Caucasus, Morocco; threatened species in Germany.

Habitat: Clearings in woods and gardens.

SIZE: 30 MM
WINGSPAN: UP TO 70 MM

THE SHORT LIFE OF AN ADULT

The adult Antlion resembles a damselfly in that it has antennae that are swollen at their tips and when at rest, its wings are folded in canopy-like fashion over its abdomen. Somewhat discreet, it only flies at dusk, camouflaging itself during daytime under vegetation. Its short life is entirely devoted to reproduction. The female carefully chooses a favorable terrain to lay her eggs. She must select a warm habitat, covered in fine sand, protected from bad weather and rich in prey, avoiding zones where other females have previously laid their eggs. She often chooses a site sheltered by rocks or large trees in order to deposit her yellow-colored, curved eggs.

Nymph after leaving its cocoon

ORIGIN OF THE NAME

The voracious appetite for ants displayed by these insect larvae is at the root of its common name. Its Latin name was, for a long time, *Formicaleo*, translated literally as "ant-lion." Since this species is common in the region of Paris (as indicated by its common name), it is called *nostras* in Latin, which means "from our neighborhood." Also known as the Doodlebug.

A VERY ACTIVE LARVAL LIFE

The predatory larva is better known than the adult. Its diabolical technique of "funnel-hunting" has been notorious since antiquity. The development of this chubby larva, measuring about 10 mm in length, lasts two years. Another one of its peculiarities: it only walks backwards! It is not unusual to find larvae in the sand at the foot of dilapidated, crumbling buildings. During summertime, the larva grows and weaves a white silk, spherical cocoon covered in grains of sand as it begins to transform itself into a nymph. Toward the end of August, the adult emerges from the cocoon at nighttime.

Larva and tunnel traps

THE "LION" AND THE ANT

An ant falls into the tunnel and tries to get out. Alerted by the vibrations, the Antlion larva assumes the position, mandibles wide open. Its twelve tiny eyes (ocelli) enable it to locate its prey. To accelerate the fall of its victim, it throws small jets of sand at it. The grains of sand get stuck under the ant's legs, causing it to slide unwittingly toward the bottom of the trap. The awful, venomous mandibles seize it. The ant dies quickly while the "lion" injects it with digestive juices that liquefy its tissues. The corpse, now devoid of its contents, will then be ejected from the hole.

THE ART OF DIGGING ONE'S HOLE

The larva's hunting technique has many advantages. Waiting at the bottom of a hole for prey to fall in does not require much expenditure of energy. In addition, the Antlion's larva, well hidden at the end of its tunnel, is protected from potential predators. The excavation of their traps occurs at night. Moving backwards, the larva begins by drawing a narrow, circular furrow of about .5 to 1.25 inches (2 to 3 cm) in diameter in the sand. It then continues in a spiral fashion, removing sand from the hole by shoveling it out with its flattened head, until a small crater is formed. The sand must be dry and the slope rigid, so that no intruder can scale it without causing a landslide. Although it is mostly sedentary, the larva can dig several such tunnels in its lifetime.

ACTUAL SIZE
Euroleon nostras
N.°
(Geoffroy in Fourcroy, 1785)
Size: 30 mm
Wingspan: up to 70 mm

Eurydema ornata

Ornate Cabbage Bug

Eurydema ornate: another coloration

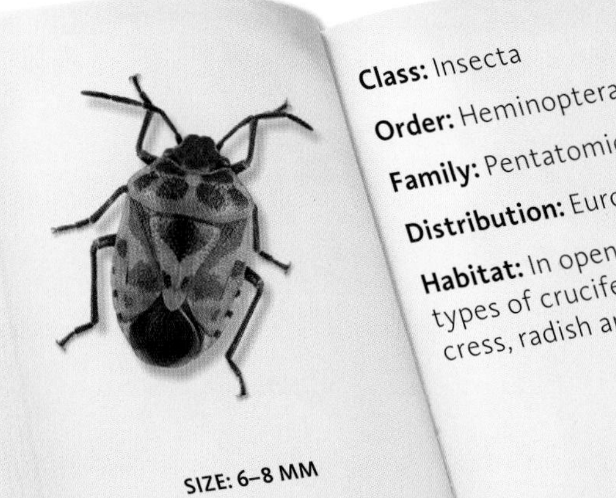

... But not exclusively

These insects are often accompanied on the same plants by another close species, the Green Cabbage Bug (*E. oleracea*), which is distinguished by its blue or green metallic reflections and embellished with red or orange (occasionally white) dots. Farmers fear the proliferation of these two species, though the risks are fairly limited, since neither one is an exclusive cabbage pest. These bugs suck the sap of a large variety of wild crucifers, and even some ombellifers. As with many other plant-eating bugs, they also occasionally behave as predators. As the larvae of syrphes or little caterpillars attack the edges of cabbage leaves, the colorful Cabbage Bugs will bite them with their rostrums.

Outspread wings

Small barrels

The adults hibernate and emerge at springtime to feed and mate. The female lays her eggs in little packets (each packet consists of a dozen eggs in two close rows) on top of leaves. The eggs look like little black barrels encircled in white, topped with a lid. After two or three weeks, at the end of spring, the juveniles hatch and immediately start feeding on sap. They must go through five molts before reaching the adult stage. Their coloration darkens as they prepare to hibernate. When they emerge the following spring, their coloration will progressively become more and more brilliant.

Origin of the name

The genus name, in Greek, means "with a large body," in reference to the shape of these bugs. The species name, *ornata,* alludes to the colorful markings on the upper surface of this insect's body. Farmers called this insect "the red cabbage-bug." However, this bug enjoys the wild as well as the cultivated varieties of crucifers.

Into cabbages ...

The Ornate Cabbage Bug can wear different costumes of variant colors (see photos). Note the antennae with five articulations (joints) that characterize this family. The scutellum covers the anterior half of the wings, which still remain visible (compare this to the Harlequin Bug on p. 132.) Adults and larvae develop on cruciferous vegetables, notably cabbage and turnips, on which they cause yellowing and rotting.

Experts in Communication

It is often the case that the male emits pheromones to attract females, rather than the opposite, as with butterflies. To answer, the female produces a "chant" at a low frequency (inaudible to humans), by vibrating the anterior segments of her abdomen. These vibrations are transmitted to the plant and received by the male as far as 6 feet (2 m) away! The male then approaches by making small hops from stem to stem, getting closer and closer, all the while emitting, in his turn, an echoing "chant." Depending on whether or not the female is receptive, she will modify her chant: it will either be a mating call or a notification of rejection!

Ornate Cabbage Bug larvae

ACTUAL SIZE
Eurydema ornata
(Linnaeus, 1758)
Size: 6–8 mm

Yellow-tailed Black Scorpion

SIZE: FROM 30 MM (MALE) TO 40 MM (FEMALE)

Class: Arachnida

Order: Scorpiones

Family: Euscorpiidae

Distribution: South of France, Italy, Spain (including Corsica, Sardinia and Majorca), Algeria and Tunisia.

Habitat: In green oak or pine forests, and cracks and crevices of old, crumbling dry walls.

ORIGIN OF THE NAME

Euscorpius signifies "true scorpion" in Greek. Its popular name is a simple translation of the species name *flavicaudis*, which means "with a yellow tail" in Latin, and refers to its venom sac.

JOURNEY OF A TRAVELING SCORPION

The scorpion, like spiders, is an arachnid: it has four pairs of legs and pedipalps that, in its case, are transformed into pincers. In spite of its strange appearance, it arouses less fascination than fear, due to the venomous stinger at the swollen posterior end (or telson) of its abdomen. It goes without saying that the sting of many types of scorpions is painful and sometimes deadly. However, the sting of the Yellow-tailed Black Scorpion is as inoffensive as that of a simple needle prick. This is clearly a good thing, since this forest species seeks refuge in old houses by finding its way into the most unexpected hiding places, such as moving trunks. That is probably how this originally Mediterranean species was able to establish permanent colonies further north including in Lyon, Bordeaux, Grenoble and even England. The Yellow-tailed Black Scorpion leaves its hiding place at dusk in search of insects, using its pincer to grab them. If the use of this appendage is not enough to subdue its prey, it will utilize its venom to immobilize it.

Sensory Combs

LOVE-MAKING PARADE

During the nocturnal mating season, between April and July, the male, after sensing the female's pheromones, is on the lookout for a partner. When they meet, the nuptial dance begins and can last several hours; this is also known as the "pariade." In a curious face-to-face ballet led by the male, pincers touching pincers, they advance and retreat, while the male shakes his tail from side-to-side and front-to-back. He jabs the female's pincer several times. With his very sensitive ventral combs (see photo), he brushes the soil to locate a spot to deposit his spermatophore, a kind of small stem that contains his seminal fluid. He recoils once in a while in order to release the spermatophore from his abdomen and to glue it to the ground, and then pulls the female to position her open genital valves over this.

GOING PIGGYBACK

Several young are born during the month of August or September. Their instinct leads them to climb up on their mother's back (see photo below). They will remain there for about ten days while awaiting their first molt, and then gradually descend to find a food source. Their development is slow: they need about 20 months and 6 molts to become adults, although several will die before reaching that stage. It has been estimated that the Yellow-tailed Black Scorpion lives for two to four years.

SURVIVAL CHAMPIONS

Scorpions have been in existence for more than 400 million years. One of the reasons for their success is their extraordinary resistance. They have endured and survived drought, fasting (an individual in captivity lived for three years without eating or drinking anything) and nuclear radiation dosages that would kill any other animal, humans included. Some desert species can even endure temperature fluctuations of more than 122 degrees Fahrenheit (50 degrees Celsius).

Mother and babies

ACTUAL SIZE

Euscorpius flavicaudis

(De Geer, 1778)

Size: From 30 mm (male) to 40 mm (female)

Common Earwig

Mother tending to her offspring

Class: Insecta

Order: Demaptera

Family: Forficulidae

Distribution: Widespread.

Habitat: Agricultural sites, hides amid leaves, under rocks or wood.

SIZE:
MALE: 14–21 MM
FEMALE: 12–17 MM

Earwig larvae

Origin of the name

Earwigs are part of the suborder Dermaptera, whose name is derived from the Greek *Derma*, "skin" and *pteron*, "wings" (their wings have a texture similar to skin). The Forficulae (another name for earwigs) are distinguished by their curved abdominal appendages known as "cerci," which resemble small forceps (*forficula* in Latin). There are lots of theories on the origin of the name "earwig." The most plausible is that it derives from the insect's habit of burrowing, which probably caused it to crawl into the ear of someone dozing in the grass or haystacks...

No shortage of food

Earwigs are nocturnal and ubiquitous; they are found in highly varied environments. During the day, they hide under rocks, under tree bark and amid flower petals. Omnivorous, they devour pretty much anything – fruits, petals and decomposing matter – and do not hesitate to eat both small insects and large, freshly dead ones. Even though earwigs can cause damage to orchards and cultivated gardens, they are also appreciated for their large consumption of aphids.

Keeping in mind the small size of their elytra, the unfolding of their wings is a complicated undertaking. The wings are folded in a fan-like fashion, then in an N-shape, before being tucked under the elytra. Earwigs do not fly much, however, and catching one in the act of this perilous, origami-like deployment is quite rare.

A mother tending to her offspring

Males begin to court females toward the very end of summer, when temperatures begin to drop. Many couplings take place. The mating ritual begins with tapping, followed by circling and counter-circling and finally mating. The female digs a hole and lays 30 to 50 eggs. Sitting at the bottom of her nest, the female fasts and becomes very combative while taking meticulous care of her eggs. She grabs the eggs with the two pairs of palpes that surround her mouth and turns them over, brushing them. Studies have shown that without such care, the eggs would become infected with fungus and consequently be destroyed.

After the eggs hatch, the mother continues to care for her young larvae until their first molt. The young gradually become emancipated and spread out from the familial shelter. The larvae look like small adults but lack wings. There are four larval stages, with adult emergence in July.

FACT OR FICTION

A very persistent legend is that earwigs pierce the tympanic membrane. They certainly use their rear pincers to defend themselves when you try to grab them, but the feeble pressure causes more fear than pain. The pincers are also used during courtship and to seize prey. The male pincers of this species are more curved than those of the female. Actually there are two types of males, those with short pincers and those with longer ones (see at right).

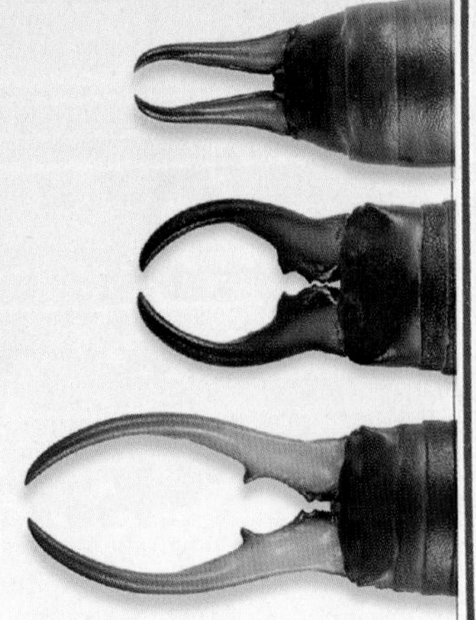

From top to bottom: female, male with short pincers, male with long pincers

ACTUAL SIZE
Forficula auricularia
Linnaeus, 1758
Size: Male 14–21 mm
Female: 12–17 mm

Southern Brown Wood Ant

Ant hill or Ne

Unlike red ants, which have a sting similar to bees, brown ants do not sting. On the other hand, they do bite with their mandibles and, if needed, spread formic acid over the wound with their abdominal glands, causing a disagreeable burning sensation.

The hunt and trade of aphids

The Southern Brown Wood Ants are omnivores. The female workers bring back small bits of prey and honeydew (sweet output from the aphids) to feed the colony. To obtain this honeydew, the ants stimulate the aphids by tapping them with their antennae. Though they support the damage caused by aphids by protecting them from tiny predators, ants are very useful, since they destroy several other noxious insects.

Nursemaid taking care of larvae

Origin of the name

The illustrious Linnaeus readily adopted the established Latin name for the ant, *formica*. The species name, rufa, means "red" in Latin. Also known as the Horse Ant.

Social insects

As with bees, ants are social insects. *Formica rufa* live in woodlands in large ant hills, which contain several queens, often in a sunny location. The dome of the large ant hill, an excellent place to capture solar heat, reaches about three feet (a meter) in height. It is built from excavated earth from subterranean tunnels, and covered with pine needles and small branches. The subterranean portion of the nest consists of a labyrinth of galleries with openings at the surface, which female workers open or close to regulate the nest's temperature. Each individual in the colony has a specific role to play that is clearly defined through a caste system.

An ant's life

In May, males and winged females crawl to the surface to mate. The future queen receives and stocks the sperm of several males, who, having lost their usefulness, die. The queen then searches for a spot to create her colony and loses her wings, which are useless for a subterranean existence. Since she cannot singlehandedly raise her larvae, she must be assisted by worker females of her species or another. In the first instance, she has no trouble in being accepted by her "assistants," but she must kill the resident queen, often a yellow ant (*Formica fusca*), before she can recruit the workers. The population of yellow ants decreases little by little until the colony solely consists of Southern Brown Wood Ants. Each egg will give rise to a worker ant or a sexual individual, depending on the quality of food received by the larvae.

The activity of the ant hill diminishes by fall and egg-laying ceases in winter, when the ants dig themselves deeper. They leave "thermal messengers" near the surface, charged with waking them up come springtime.

Defensive posturing

ANTS BY NUMBERS

- 100,000 ants: the number of individuals in some nests.
- 7 ants/m^2: these are the most densely populated and numerous insects on earth.
- 5 ounces (140 gm): the estimated quantity of honeydew brought to the nest every day.
- 1 billion: the number of ants on the planet.
- 100 million years: the age of the oldest dated ant fossil.
- About 10 years: the life expectancy of the queen.
- 60 times its weight: the weight that one ant can pull.

Formica rufa

Linnaeus, 1761

Size: Queen and male: 9–11 mm

Female worker: 4–9 mm

ACTUAL SIZE

Common Pondskater

Pondskater larva

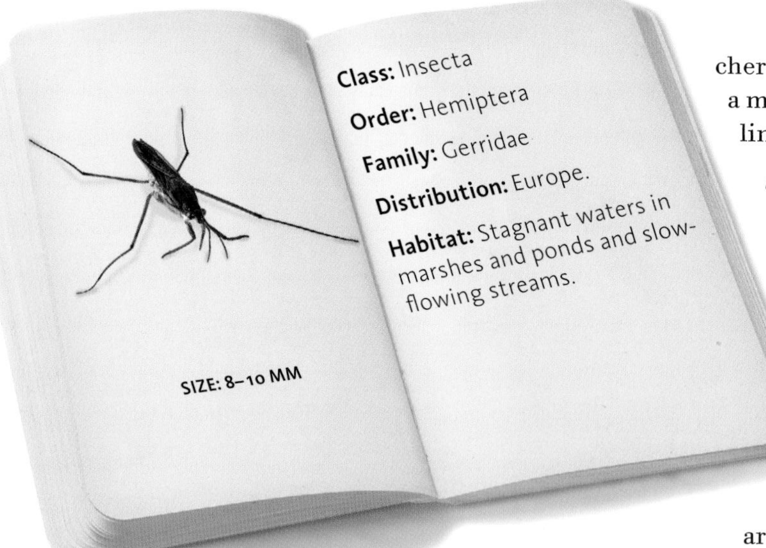

Origin of the Name

The name *Gerris* is derived from the Latin *gerres*, a small unidentified fish cited by the Roman author Pliny. It is an allusion to this insect's habitat. Likewise, the name lacustris is derived from *lacus* or "lake."

Pondskaters

Commonly known as "water spiders," "scissors" or "pondskaters," the Gerridae and similar bugs have very elongated bodies and long, thin legs. In this particular species the male has a black abdomen and the female a yellowish one, which renders her less visible to aquatic predators than her partner. The Pondskater runs across the water surface in a jumpy fashion (that is why it is also called the "Jesus bug!") Its long, spread-out legs enable it to apportion its weight over the water, to the point that it appears to glide across the surface. Moreover, its feet are aligned horizontally and equipped with hair-like bristles coated with a hydrophobic (water-repelling) substance, which is secreted from a gland near its mouth. The front legs (often hidden) are used in predation, the middle legs for rowing and the hind ones as rudders and for stability. While studying the Pondskater's movements, researchers have recently discovered that the middle legs create a minute swirl under the water, which acts like a trampoline, propelling the bug forward.

Still Waters Predator

Pondskaters live in marshes, ponds and other bodies of water, as well as calm areas in streams. Quite often they gather in groups (see photo), forming symmetrical patterns that often attract the attention of people strolling nearby. Pondskaters feed on insects and other small invertebrates, dead or alive, which have fallen into the water. Due to their sensitive legs, which can detect vibrations, pondskaters can spot prey from afar. They pounce on the prey, seize it firmly with their front appendages, and then pierce it with their rostrum to aspirate its contents. Sometimes, they also practice cannibalism.

Two Generations per Year

After hibernating over the winter, the adults reproduce in the spring. As with most bugs, mating is rather brutal; the male, which is smaller, penetrates the female anywhere – the abdomen, the thorax, even her head! The female body has the ability to concentrate the male's semen and direct it to her reproductive organs. In order to prevent another male from mating, he stays attached to the female in this mating position and gets carried around like this for quite some time. The eggs are deposited in clusters on water plants and are held together with a gelatinous material. To breach the egg envelope, the larvae use a type of cutting blade located between their eyes and then they swim to the surface. The juveniles are similar in appearance to adults, but smaller and squatter. Although some juveniles have wings, a proportion of them does not. The second generation of adults reproduces at the beginning of summer. Once their offspring become adults, the ones with wings fly some distance from the water to hibernate under vegetation on dry ground.

Devouring prey

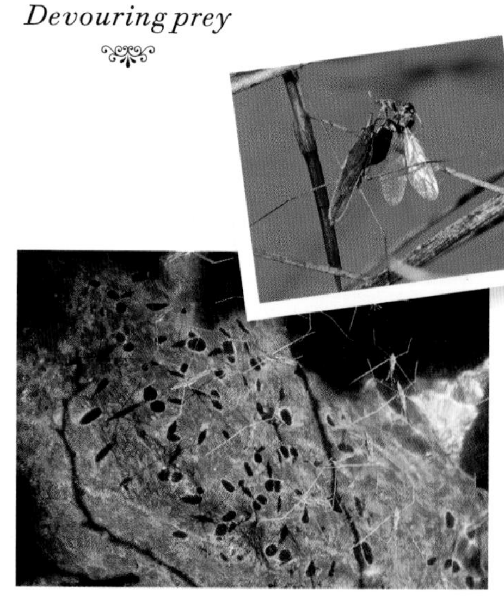

A group of pondskaters

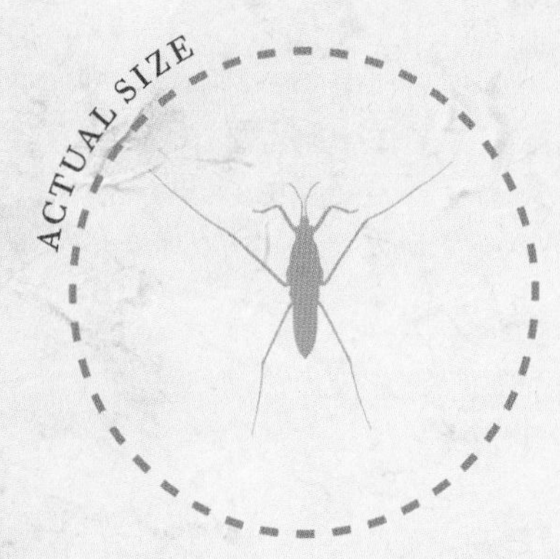

Gerris lacustris

(Linnaeus, 1758)

Size: 8–10 mm

Pill Millipede

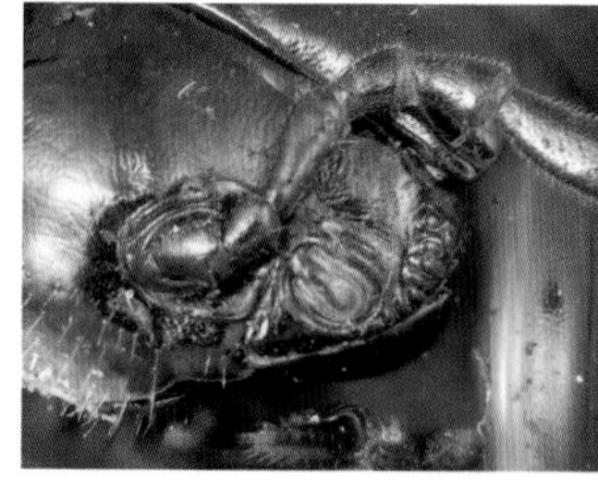

The underside of the Pill Millipede. Count the legs!

ORIGIN OF THE NAME

The name Glomeris comes from the Latin *Glomerare*, which means, "to fold into a ball." The species name refers to the clear markings that delineate each ring of their body. J.M. Demange, a specialist in Myriapodes, has amusingly described these clear delineations as "pouches on a too-tight costume."

A SHORTENED MILLIPEDE

Attention! The Woodlouse (p. 186) and the Pill Millipede (Glomeris), though found in the same kind of habitat, are in no way related to one another. The Cloporte is a crustacean "misplaced" on solid ground while the Glomeris is a millipede with a particularly short body. If you flip a Glomeris over, you will notice that each body segment has two pairs of legs (for a total of 17 pairs of legs in the adult). The Cloportes have only one pair of legs per segment, and seven pairs in all.

GOOD IDEAS ALWAYS GET STOLEN

Pill Millipedes only appear at night. During the daytime, they can be surprised, alone or in groups, under a piece of wood or under leaves on the forest floor. They feed on decomposing plant debris (leaves, wood, etc.). Apart from its solid and rounded carapace, it also has a secret weapon to protect itself against predators: it produces deterrent substances that are very indigestible, stored in special glands. The red robin, a very cunning bird, uses these secretions to good advantage by rubbing the Glomeris on its feathers to rid itself of burdensome parasites.

APHRODISIAC ODORS

Reproduction occurs in springtime and in the fall. The male attracts his beloved by emitting pheromones. Millipedes perceive these odiferous signals through a specialized horseshoe-shaped organ, located at the base of the antennae (see above image). During mating, the male emits other pheromones that result in the female's immobilization. In Diplopods, the genital apparatus of the male is located behind the second pair of legs and the copulatory legs (called "gonopods"), at the level of the seventh ring or annulus. "Mister" Glomeris is the exception to this rule, as his gonopods are located at the body's extremity. Consequently, he has to behave like a real acrobat to be able to fertilize his female!

The female lays 70 to 80 quite large eggs (1 mm in size), each one protected by a coating of regurgitated mud. From this emerges a membrane-enveloped embryo that, after a few days, transforms itself into a larva. The larvae resemble the adults, but their bodies have fewer segments and their legs augment through the successive molts that lead to adulthood.

THE PERFECT SPHERE

As soon as a hedgehog, armadillo or other predators appear at the scene, the Pill Millipede rounds itself into a ball to protect itself against the aggressor. Its annuli or rings adjust themselves perfectly, leaving no space in between, and the head is folded inside its legs while balling up. In this way, the last annulus or ring links to the first one, thereby protecting the soft and vulnerable ventral body parts. This exceptional protective mechanism may explain why these animals have such a long lifespan: up to 11 years!

Glomeris marginata

(Villers, 1789)

Size: 7–20 mm

Lemon Butterfly

The Lemon Butterfly at rest

Class: Insecta

Order: Lepidoptera

Family: Pieridae

Distribution: From Europe to North Africa to Mongolia.

Habitat: In open spaces or woodland clearings; the caterpillar on the buckthorn or the alder buckthorn.

WINGSPAN: 55–65 MM

The female is less colorful

FROSTED LEMON

The Lemon Butterfly is one of the rare diurnal butterflies with enough resistance to be able to hibernate in its adult stage in colder areas. It usually hatches in July, but it only flies for a brief period of time before entering aestivation, or summering, in the heart of summer. It shows itself in the fall before entering a long winter diapause in a sheltered area (barns, hollowed trees, ivy leaves, etc.). It reappears with the warm sunny days at winter's end or at the very onset of spring. The males have a tendency to awake sooner than the females, as soon as a ray of sunshine gives them enough energy to feed at the willow's catkins. We have occasionally encountered frozen specimens who were slowly coming to life under the warmth of the sun! This ability to withstand both harsh summers and cold winters explains the exceptional long life of this butterfly in its adult state.

ORIGIN OF THE NAME

The yellow color of the males is what gave this insect its common name. Its scientific names are quite full of imagery: *Gonepteryx* means "angular wing" (from the Greek *gonia*, "angle," and *pteryx*, "wing"), while *rhamni* simply refers to its host plant (Rhamnus is the genus name of the buckthorn and the alder buckthorn).

MASTERS AT CAMOUFLAGE

This butterfly is a cousin of the Large White Cabbage Butterfly (p. 200). It is the most remarkable predictor of springtime: as of February it can fly on clear, warm sunny days. With its yellow coloration (paler in the females, see above) and diurnal nonchalant flight pattern, it is easy to spot as it skims the tops of hedges, looking for the willow catkins from which it loves to gather pollen and nectar. At rest on vegetation, this insect is practically invisible. With folded wings and the tree-like colors on their underside (yellow or greenish with spots of rust), they look very much like leaves.

FROM THE EGG TO THE LEMON

In springtime, the Lemon Butterfly executes its nuptial parade. The female spreads her odiferous scales to attract the male and offer him her abdomen. The fused eggs, of a yellow-green color, are laid exclusively on the buckthorn or the alder buckthorn leaves. The green caterpillars, dotted in black, and marked with clear lateral lines, feed on the leaves of the host bush. Metamorphosis occurs at the beginning of summer. The chrysalis, hanging from a branch of the host plant, mimics a folded leaf. The butterflies hatch soon after and feed for a while on flowering hay and flowers at the edge of the forest, before undertaking their summer rest.

SOUTHERN LEMON BUTTERFLIES

Among species of the genus *Gonepteryx*, the Lemon Butterfly of Provence (*Gonepteryx cleopatra*), is certainly the most beautiful. The male, in particular, has forewings broadly sketched with bright orange, which creates a very handsome effect. As its name indicates, the Lemon Butterfly of Provence has a more southern distribution, and is found throughout in the entire Mediterranean region.

Caterpillar and chrysalis

ACTUAL SIZE

Gonepteryx rhamni

(Linnaeus, 1758)

Wingspan: 55–65 mm

Italian Stink Bug

SIZE: 9–11 MM

Class: Insecta

Order: Heminoptera

Family: Pentatomidae

Distribution: Europe (rare or absent in the North).

Habitat: Often in large numbers on wild carrots (Queen Anne's Lace) and other ombelliferous plants.

Graphosoma semipunctatum

Origin of the name

The scientific name refers to the ornamentation of this beautiful insect: *Graphosoma* means "body with inscriptions" in Greek. These amazing markings have also given it many other common names: Harlequin Bug, Lined Pentatome or Italian Graphosome. The word "pentatome" comes from the family name Pentatomidae, originating from the Greek meaning "in five parts;" in this group, the antennae are composed of five joints (compared to four in other bugs). These pentatomes are also called "Bugs with Shields" in reference to the massive form of the triangular scutellum that largely covers the wings.

Expert in foraging

The Stink Bug is one of the most beautiful bugs, with a bright red body striped longitudinally with black bands on the back, and punctuated by black on the stomach and legs. A slightly different species, *G. lineatum*, with entirely red legs, is found in North Africa. Long considered synonymous, these two species are in fact quite distinct from one another. The Stink Bug prefers very sunny spots well endowed with flowering low-growing plants (as found in meadows, embankments, etc.). This insect is observed mostly from May to September on ombelliferous flowers (fennel, wild carrot, angelica, etc.), where it is often found in great numbers. Its coloring reminds its predators that it is toxic. Similar to many other beetles, it has glands on the side of its thorax that exude an awful-smelling liquid when the insect is seized. The punctuated Graphosome (*Graphosoma semipunctatum*) is a very close species, distinguished by discontinuous stripes on the pronotum (to the left).

A "close" couple

Reproduction occurs in springtime. Following a brief dance, during which the two partners touch one another with their antennae, mating occurs very much like that of all these types of insects: head-to-foot. The couple remains attached this way for many hours, often many days. This is a sure way for the male to assure himself of the exclusivity of his female and his progeny! The fertilized female lays her eggs in small packets on the leaves of an ombelliferous plant. The larvae are born a few weeks later. They will then have to go through five successive molts before becoming adults.

BEWARE THE RED AND THE BLACK

It is interesting to note that a fair number of insect species have adopted the red and black coloration, including coleopterans like the Ladybug (p. 88), the Six-Spot Burnet Moth (p. 248) and other bugs such as the Firebug (p. 208). All these insects move about openly without concern for predators – their vibrant coloration is said to be "aposematic" and sends a strong message: "Beware! I bite" or, more likely, "I taste bad!" or, "I am toxic!"

Birth of larvae

N. *Graphosoma italicus*

(Müller, 1766)

Size: 9–11 mm

Gryllotalpa gryllotalpa

European Mole Cricket

Elytra of the female (left) and male (right)

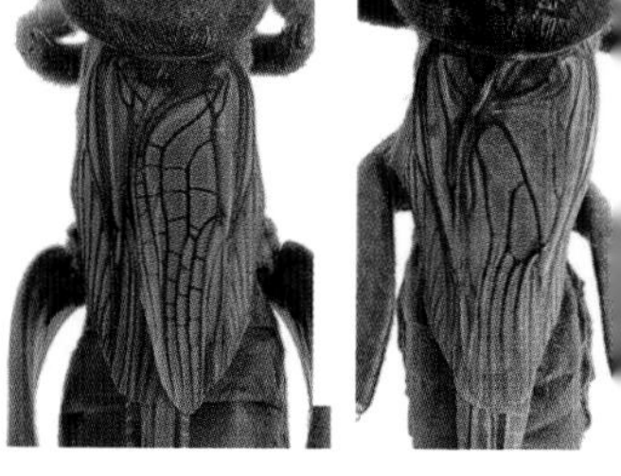

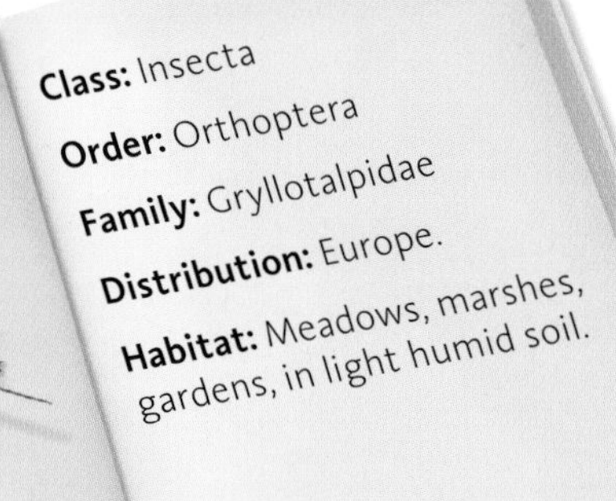

SIZE: FROM 35–50 MM

The sound is amplified by the underground tunnels, allowing the females to detect potential partners from afar. The female mates with the best singer and builds a chamber of compacted earth the size of a fist, in which she deposits 200 to 300 eggs. Lest gardeners cringe at the thought of an invasion, very few of these offspring attain adulthood!

As with Earwigs (p. 122), the mother stands guard over her nest until the eggs hatch and the small larvae leave the nest and burrow their way out. The juveniles spend the first winter buried underground. The following spring they grow in size and molt one final time, just before winter. They spend another cold season buried in soil and do not begin reproduction until spring of the third year.

Mole Cricket larva

Origin of the name

This cricket likes to live among vegetables. It is known by the evocative name of Mole Cricket (its scientific name *Gryllotalpa* means the same thing in Latin). Because of its massive pronotum and strong anterior limbs that resemble claws, it was formerly called the "ground crayfish."

Living underground

The mole cricket is well adapted to life underground: its claw-like anterior appendages are perfect for moving earth, and its strong, tapered body allows it to move with ease. It likes soft, damp soils at the edge of marshes or flowing waters, but also grasslands, gardens and manure. Males are distinguished from females by their wings: the elytral cells are larger and fewer in number. Despite their appearance, mole crickets can fly by deploying membranous wings, which are folded along the abdomen.

Love songs in the tunnels

The soft song of the males can be heard in the spring, the result of vibrating wings that emit a chirping sound as they overlap.

It has become rare today

The mole cricket is omnivorous, cruising tunnels in search of worms and larvae to consume or a root to nibble on. Because of its exaggerated reputation as a pest it has been badly treated. Today, due to insecticides, lack of manure and above all, agricultural machinery and other automated cultivating devices, the mole cricket has become rare.

THE INSECT MOLE

The front appendages of the mole cricket bear a striking resemblance to those of moles. This phenomenon, well know to biologists, is called "convergent evolution." This refers to the fact that unrelated organisms living in similar environments have evolved similar morphological characteristics. Note that the mole cricket's pronotum is coated with a fine hair-like covering, soft to the touch, much like miniscule fur.

ACTUAL SIZE
Gryllotalpa gryllotalpa
(Linnaeus 1758)
Size: 35–50 mm

GRYLLUS CAMPESTRIS

Field Cricket

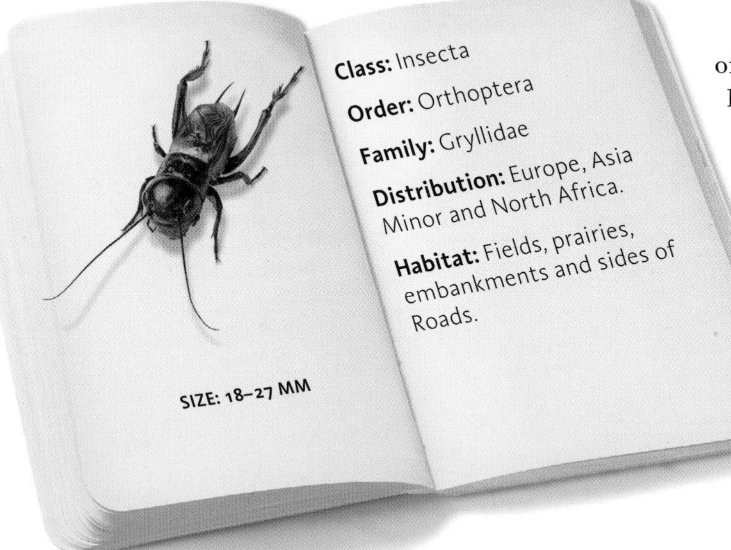

ORIGIN OF THE NAME

The scientific name derives from the Greek *gryllos*, used already to designate this insect. The species name, *campestris* ("of the fields"), originates from the Latin in reference to its natural habitat.

THE CRICKET'S VIOLIN

As soon as April rolls around, the fields resound with crickets. At this time of year, the other orthopterans (grasshoppers) are still at the egg or larval stage. They will not take over until much later in the season. The "cri-cri" (stridulation or chirring) sound, slightly repetitive, is emitted by the male cricket. The male sits at the entrance of his burrow to "sing" by rubbing his elytra against one another. The edge of the right elytra has a regular rough line (the stridulation rasp), which rubs against the thickened border (the "harp") of the left elytra. The male's "song" is primarily designed to attract and seduce the female, but also serves as an early warning of its presence to rival suitors.

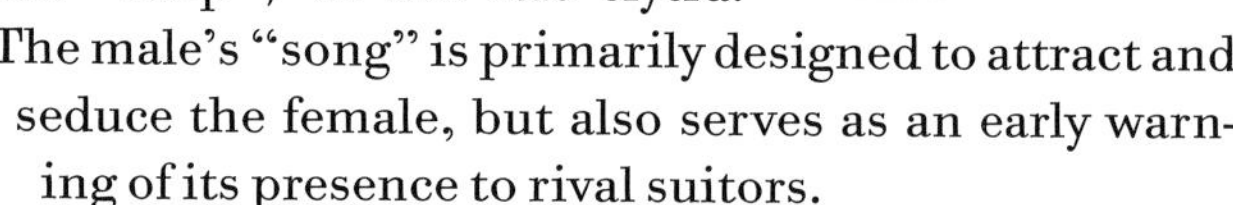

Male surprised in full "song"

THE MUSICAL EAR OF THE FEMALES

The female directs her "ear" toward the stridulant virtuoso posed at the entrance of his lair. As with grasshoppers, her auditory organs (the tympani) are located in the tibia of her forelegs. As soon as a female approaches, the male modifies his "love song" to seduce her and rotates around her and taps her with his antennae. If she is receptive, he introduces a kind of vial, the spermatophore, into her genital opening. The spermatophore will subsequently swell and expel its spermatozoids. In April or May the fertilized female buries her ovipositor in the soil and lays her eggs. By July, the Field Cricket's song is no longer heard, since all the adults have died.

The young nymphs emerge quickly in July and August and crawl out of the burrow. They grow rapidly by successive molts and, as omnivores, feed mostly on plants, but also small insects. By the time fall comes, the larvae seek out an old burrow, or build one of their own in order to hibernate during the last larval phase. In the following springtime, they awaken; go through their final molt and go on to live their lives as adults.

SABER-CARRIER WITH LONG ANTENNAE

Despite their squat appearance and their mediocre hopping ability, crickets closely resemble grasshoppers. To understand this similarity, one needs only observe their long, thin antennae, and the female laying organs (ovipositors) in the form of a saber's blade. In crickets, the females' antennae and the ovipositors are much shorter. Other than the presence of an ovipositor, the female cricket is distinguished form the male by the veination of the elytra: hers is more regular and less modified than that of the male.

CRICKET GAMES

In certain regions of China, cricket fighting has replaced cock fighting. Not so long ago, country children took part in a less cruel game: cricket-chasing. This "tradition" is slowly disappearing. It is not very difficult to find a burrow in vegetation or under a tuft of grass. At the slightest warning, the "tenant" quickly silences itself and dives down to the bottom of its burrow. All one needs is a blade of grass to tease the insect out.

Field Cricket larva (left) and adult

ACTUAL SIZE
Gryllus campestris
Linnaeus, 1758
Size:
18–27 mm

Gyrinus substriatus

Common Whirligig Beetle

These beetles are very gregarious

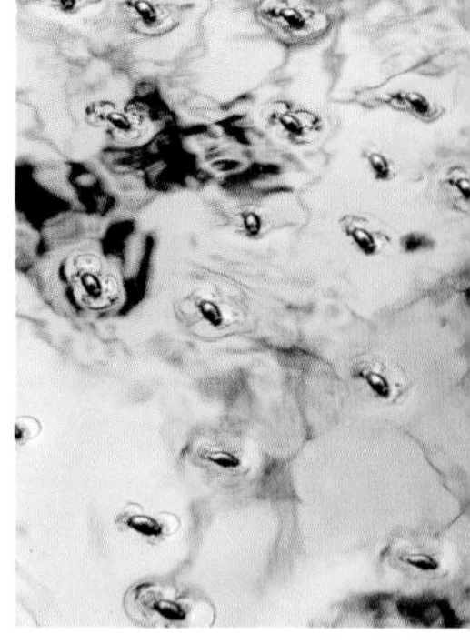

Class: Insecta

Order: Coleoptera

Family: Gyrinidae

Distribution: Western and Meridian Europe.

Habitat: Ponds and rivers.

SIZE: 5–7 MM

Origin of the name

The Whirligig Beetle owes its name to its unusual behavior. In Latin, *Gyrinus* literally means "one that spins around oneself." Its species name, *substriatus,* refers to the reduced striations on its elytra. The species illustrated here has long been confused with the *Gyrinus natator* of Linnaeus, which is found in more northern regions.

Whirling dervishes

It is during summertime that these beetles are most evident on the surfaces of lakes and ponds; they gather together and perform their strange, dizzying behavior! If we approach this band of "whirling dervishes," they go crazy and swim faster and faster, covering distances of up to 20 inches (50 cm) per second and passing each other without colliding. Then suddenly they dive into the water and grasp onto submerged plants. We must wait perfectly still for a few minutes to see them reappear and start their nautical ballet one more time.

The speedboats of ponds

Whirligig Beetles feed on tiny insects, dead or alive, that fall in the water. They spread their two little stocky antennae on the water's surface to detect the vibrations that signal their prey's presence.

The beetle's forelegs are used mainly to grasp its prey, while its two other pairs of legs are designed for swimming. The latter are equipped with fine bristles, which it deploys to steady itself on the water during propulsion. The bristles fold back along the legs when the insect brings its legs close back to its body. The legs tuck into small pockets under the "hull" of this miniature speedboat, facilitating a sliding motion. These beetles are denser than water, but they float thanks to the surface tension created by an oily, aromatic secretion. A dead Whirligig Beetle sinks to the bottom of the pond or lake.

Natural history

The Whirligig Beetle spends the off-season hiding under vegetation or under rocks on embankments. It reappears, often quite early, during the first warm sunny days in spring.

When the mating season arrives, the Whirligig Beetle flies off in search of new water location. After mating, the female lays about thirty eggs on floating debris. Miniscule white elongated larvae with short legs are born soon after. They are dotted with 10 "tracheobranchiae" edged with fine hairs, which assist with respiration as well as with navigating through water. The larvae continue to grow at the bottom of water, feeding on tiny insects, which they kill and digest in an extra-oral manner, like the Great Water Beetle (p. 110). Approximately one month after egg-laying, the larvae construct a small loge, in which they will metamorphose into nymphs.

Two or four eyes?

The Whirligig Beetle faces threats that come both from the air and from the water. Its tactic: double, simultaneous vision, both aquatic and aerial! Contrary to all appearances (see photo), the Whirligig Beetle only has two eyes, just like all other coleopterans. However, each eye is divided into two distinct parts, one located on top of the head in the air, the other located under the head and submerged in the water.

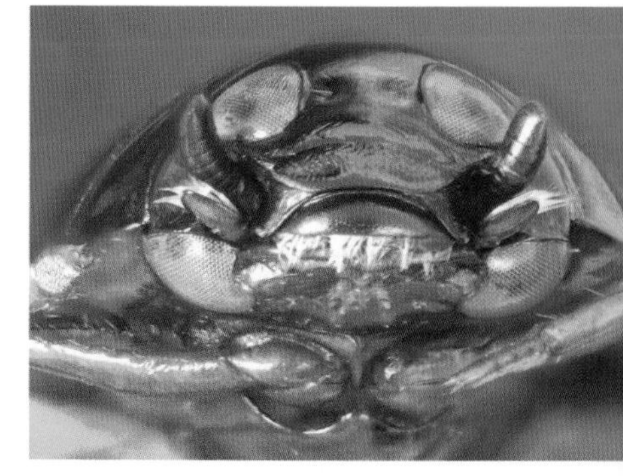

Gyrinus substriatus

Stephens, 1829

Size: 5–7 mm

HAEMATOPOTA PLUVIALIS

Rainy Day Horse Fly

Another species of horsefly

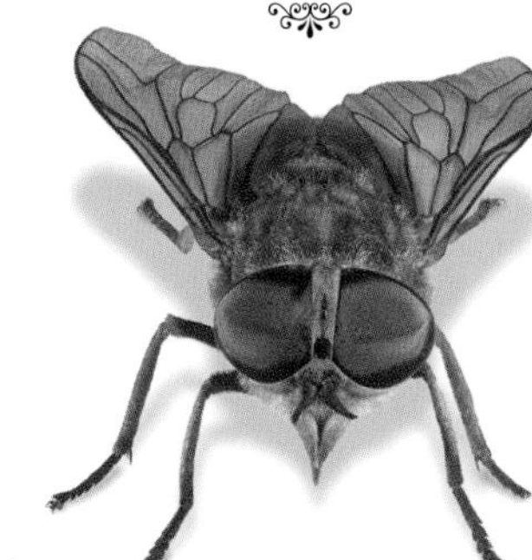

SIZE: 5–7 MM

Class: Insecta

Order: Diptera

Family: Tabanidae

Distribution: Europe and North Africa.

Habitat: Ubiquitous, the adult moves a lot; larvae are found in humid areas.

Origin of the name

The species name *Haematopota* was coined in 1893, by Meigen. In Greek, *animatos* means "blood" and *potès* designates the drinker. One would be hard pressed to find a more appropriate name! *Pluivialis*, in Latin "from the rains," alludes to the fact that this insect will bite even under humid conditions.

No respite, even when it rains!

The Rainy Day Horsefly is a very common species. Unlike other species, it remains active during cloudy and rainy days, and does not hesitate to attack humans.

The female lays about one hundred little, elongated black eggs that are arranged, three-tiered, in inclined mounds on aquatic plants. In one to three weeks, tiny maggots will emerge and, during one year, go on to become the fearsome predators of other larval insects in the soil or mud. By springtime, the transformation into pupa occurs on drier grounds, close to the water's surface. The pupa has a little crown of teeth that helps it crawl to the surface of the soil to facilitate the adult's emergence. It is our good fortune that the adult often falls prey to spiders, damselflies, predatory flies (Asilidae) and several birds.

Only the females bite

Just like the mosquito, it is the female Horsefly that bites, while the males are inoffensive gatherers of pollen and nectar from flowers. The female mouth parts develop into long, pointed, sharp blades. This double pair of "surgical knives" enables her to pierce even the toughest of leather-like skins, after which a small, enlarged "sponge" (the labium) helps capture the blood flowing from the bite.

After targeting a potential "donor," be it a bovine or a horse, the female lands on a spot least subject to the fly-swatting body movements of her intended prey. When the pain of the sting is felt by the hapless victim, it is often too late and the Horsefly takes off with her little mouthful of blood. This blood is crucial for the development of her eggs. Such attacks, often involving large numbers of Horseflies, can cause quite a disruption in a herd of animals.

The ancient horseflies

The Tabinidae are a very old Dipteran Family, first appearing some 230 million years ago. Knowledge of these insects goes back a long time in human history. The first known written mention of them appears in the Old Testament, when an invasion of Horseflies constituted the third plague of Egypt

Packet of eggs on a leaf

YOU KNOW THAT YOU HAVE GORGEOUS EYES!

Very often, in the world of insects, eye colors are "matched" to those of the body. When an animal is lucky enough to have a coloration that matches that of its environment, the eyes are colorless. This is not the case in Horseflies: their enormous, colorful eyes contrast markedly with their often faded body coloring. The eyes of the Rainy Day Horsefly are magnificently striped. These eyes (adjacent in males and separated in females) are composed of millions of simple eyes.

Rainy Day Horsefly ogles the camera

ACTUAL SIZE

Haematopota pluvialis

(Linnaeus, 1758)

Size: 5–7 mm

Hispa atra

Black Leaf Beetle

SIZE: 2.5–3.7 MM

Class: Insecta

Order: Coleoptera

Family: Chrysomelidae

Distribution: From eastern and central Europe to northern China; North Africa.

Habitat: Open terrain, meadows, fallow lands, marshes, sometimes in clearings.

Hispa larva

Origin of the name

This Black Leaf Beetle (also called Hirsute Hispa, or Hispa) does not really have a common name. The biologist Geoffroy, who discovered this insect, prudently called it the "The Black Chestnut," referring to its similarity to the chestnut's spiny outer seed covering. This name never found wide acceptance, however. The beetle's scientific name comes from the Latin *hispidus*, meaning "bristling," and *atra*, "black." It is true that with its spiny elytra, pronotum and even its forelegs and antennae, this beetle looks like a small hedgehog!

Misleading appearance

With its almost exocentric appearance, it's hard to realize that this black beetle is distantly related to the Colorado Potato Beetle (p. 154) or the Rosemary Beetle (p. 80)! However, that is indeed the case, and the black Hispa is part of a distinct subfamily. This insect is quite common in France, though it is only seen on an irregular basis. Its geographic range seems to be extremely wide, although some authors are not convinced it is that widespread; it seems more likely that the specimens collected in France, the Caucasus, central Asia and Mongolia are closely related but indeed different species.

Mating

Discretion is assured

The black Hispa beetle is unobtrusive in the meadows and fallow lands it inhabits. Its small size makes it go unnoticed among the grasses it inhabits. Like many other insects, it falls to the ground if it is threatened or if a curious observer approaches. It disappears completely in the tangle of roots and the drab soil of meadows. The simplest way to observe the adults is to lie down in tall grass and examine the shafts of vegetation around you. With a bit of luck, the bristly little insects will dance before your eyes!

Small juvenile larvae

In the spring, the female inserts individual eggs on the edges of leaves of grasses and covers them with mucus, which hardens. The young Hispa live hidden in the host plants, especially on couch grass and in pastures. They are small, translucent, very flattened and have three pairs of short legs. They travel in tunnels at the base of plants, where they transform themselves into chocolate brown nymphs with posterior spines. The adults emerge in the fall (there is only one generation annually) and pass the winter underground.

AN EXOTIC INSECT LOST IN FRANCE

The ruffled appearance of this black beetle is common among several species of the subfamily Hispinae, and makes one think immediately of an exotic insect. It is difficult to picture such exuberance in a moderate climate. The majority of the 3,000 species in this family live in the tropics, with only four in Western Europe, two of which are found in France. Another French species, *Dicladispa testacea* (see photo), is found in the Mediterranean regions and Corsica, living on its host plant Cistus.

ACTUAL SIZE

N.° Hispa atra

Linnaeus, 1767

Size:

2.5–3.7 mm

Hoplia coerulea

Blue Scarab Beetle

Origin of the name

The genus name *Hoplia*, described by Illiger in 1803, signifies "claw" in Greek, and refers to the long claw of the tarsus. The name of the species, *coerulea,* "azure" in Latin, refers to the beautiful metallic blue color of the male.

Small, azure-blue beetle

The Blue Scarab Beetle resembles other small beetles with metallic colors. This beetle is one of the rarest coleopterans that display both scaly and colorful body armor. The structurally complex scales are composed of several layers of chitin and air, which diffract light very much like a mirror. However, the female (photo) is much less discreet in her coloration: her translucent scales permit the transmission of the underlying tegument color, so that she appears brown with silvery reflections.

The ladies play hard-to-get

This beautiful insect inhabits humid areas during summer, near flowing rivers where the roots of low-growing plants provide nourishment to the larvae. The adults, on the other hand, feed on the flowers. The Blue Scarab Beetle can be abundant some years and hard to find in others. The males swarm in bright sunlight. The females, on the other hand, are more discrete: rarely visible, the ladies come out solely for mating. The deed usually occurs at noontime when the sun is at its highest point. The males take their place at the edges of waterways, grasping onto the highest weeds with their forelegs, while lifting their posterior ones. The females need only leave their shelters and choose among the many suitors.

Completing this "outfitting"

The genus *Hoplia* is widespread in the entire Northern Hemisphere, from the United States to Japan, with about 300 species. If the Blue Scarab Beetle is indisputably the most remarkable, the Silver Scarab Beetle (*Hoplia argentea*) is one of the most widespread. It is seen from May to July in the neighborhood of rivers, and often on wild roses. The even more widespread Chalky Scarab Beetle (*Hoplia farinose*) is brown, with a slight covering of clear, loose scales mixed with yellowish hairs.

A jewel among a nest of flowers

During the 19th century, the fashion favored this species, which was generally widespread at the time. The colorful beetles were used in flower arrangements and also to create elegant personal ornaments (collars, broaches, etc). People would collect them by the thousands in the Loire Valley, to sell to florists. Times have changed and the numbers have considerably diminished, even though one can still occasionally find groups of males hanging from vegetation along waterways. However, this is episodic and localized.

The female

Males waiting for the females

Hoplia coerulea

(Drury, 1773)

Size: 8–10.5 mm

Peacock Butterfly

Emergence

Peacock butterfly larva

Origin of the name

The scientific name stems from a double reference to a famous Greek legend. Inachos, the river god and son of Oceanos and Tethys, was the father of Io, priestess of Hera, Zeus' wife. Seduced by Zeus, Io suffered the jealousy of Hera, who transformed her into a cow and appointed Argo, a monstrous herdsman with 100 eyes, to guard her. Zeus had Argo killed, and in order to honor him, Hera had his eyes transferred to the feathers of her favorite bird, the peacock. This butterfly's common name obviously refers to the eyelike spots on its wings, reminiscent of those on a peacock's tail.

Day and night

The Peacock Butterfly is one of the most common in the region, even found in proximity to inhabited areas. The beautiful ornamentation on the upper surface of the wings – identical in both males and females – is quite striking: crimson red with a velvety aspect, and large ocelli ringed in yellow and highlighted with metallic blue. The under surface is dull with a sootlike coloration. The reason for this difference is quite simple. When it lands on a branch at night or in a dark corner to hibernate, it folds its wings, thereby blending in with its surroundings. When disturbed, this butterfly quickly spreads its wings, revealing its big "eyes" as a means to distract its predators and allow it time to flee.

Exceptionally long lived

When adding up the many months the Peacock Butterfly lives as an adult (about 10!), we find that it is among the most long-lived of known butterflies. This lengthy lifespan is possible because adult Peacocks hibernate. As fall approaches, the butterfly hides in barns, woodpiles or dead trees – and even in our homes. This can prove to be fatal, however, if the heat in the house causes the butterfly to prematurely awaken. When springtime approaches and the first rays of sunshine strike the willow's catkins, she feeds on nectar before mating and depositing her eggs.

An initially gregarious lifestyle

In April to May, the female lays her eggs in clusters under hop and dandelion leaves. The eggs are a light green color with white stripes. After hatching, the small yellowish caterpillars begin life in groups sheltered in a common web. In its final stage, the caterpillar – now black with white spines – measures about 1.5 inches (4 cm) in length. After the last molt, the caterpillar leaves its host plant to transform itself into a yellow-green chrysalis (see below), hanging from a tree branch or trunk.

It's raining blood

From the Middle Ages to not that long ago, a variety of authors have written about the disquieting phenomenon of "blood" raining down from the skies. Is is a sign from God, a sinister omen or an act of the devil? Many frightened souls have let their imaginations run wild trying to explain such astonishing tales. There is nothing supernatural, however, about finding some clothes left under a tree to be covered with blood. As with the Red Admiral Butterfly (p. 240), this may be the result of a viscous red liquid, the meconium, being expelled from the rectum of Peacock Butterflies when they hatch. So a peak of emergence can thus be misinterpreted as a stream of blood!

Transformation into a chrysalis

ACTUAL SIZE

Inachis io

(Linnaeus, 1758)

Wingspan: 55–60 mm

European Spruce Bark Beetle

Origin of the name

In Greek, Ips designates a worm that gnaws on wood, and the word *scolyte* (from the family name Scotylidae) comes from another Greek word (*skolès*), which also means "worm." *Typographus* (from the Greek, literally signifying "to write imprints") is a classical composition and printing method used for printing relief characters. Obviously, our "typographical" insect does not take itself for a Gutenberg. Its name arises from the magnificent stylistic ornamentation sculpted by the females and their offspring in spruce tree trunks. This design is so characteristic of each species that it constitutes a near-infallible identification feature.

A methodical enterprise

Tree tunnels are dug according to a very precise plan. The male enters the bark first, at an angle perpendicular to the tree's axis. Since he is polygamous, he digs a small chamber designed to accommodate several females. The females, which are attracted by trail of pheromones, soon join the male in his lair and mating begins. Each female then digs a longitudinal tunnel destined to become an egg-laying corridor, and marks her tunnel with dozens of small, regular chambers in which she will deposit one egg, and close the entrance with some compacted sawdust. By the process of backing out of the tunnel, the male in fact cleans it by brushing the inner surfaces with his bristle hairs, thus forcing the sawdust out. The truncated form of its elytra is well suited for this bulldozing work.

After hatching, the larvae feed by digging their own tunnels, perpendicular to that of their mother's. They let the bottom of the tunnel become obstructed by their excrement. As they progress in this manner, the tunnels become larger and larger as they grow in size. At the end of this growth period, they stop working and molt on site to become nymphs. Once adults, the Spruce Bark Beetles forage an exit hole and leave in search of another similar tree trunk to establish themselves.

Strength in numbers

The Spruce Bark Beetle generally only attacks weakened spruce trees (previously attacked by caterpillars or fungi) or those that have recently fallen. In fact, a healthy tree efficiently defends itself by inundating its aggressor with sap, thereby gluing it in its resin. If the supply of bark is plentiful at the time the beetles spread, these insects reproduce rapidly. When they begin to proliferate, they even attack healthy trees. With their bark and cambium (the layer between the bark and the wood) full of tunnels, these giant trees can't help but die.

REASONABLE RAVAGERS

The Spruce Bark Beetle has the reputation, deservedly at times, of a being a fearsome ravager to battle against. These ravages, however, only take place after extensive felling of a similar species of trees in the forest. Intensive exploitation and forest restocking with monoculture trees also promotes the proliferation of these beetles. In diversified forests, the impact of these insects is quite modest and, through eliminating sick trees, they actually play a significant ecological role.

Tunnels "drawn" on spruce bark

Bark beetle larva (left) and nymphs (right)

Ips typographus
(Linnaeus, 1758)
Size: 4–6 mm
ACTUAL SIZE

Ixodes ricinus

European Castor Bean Tick

Copulation: the female is on top.

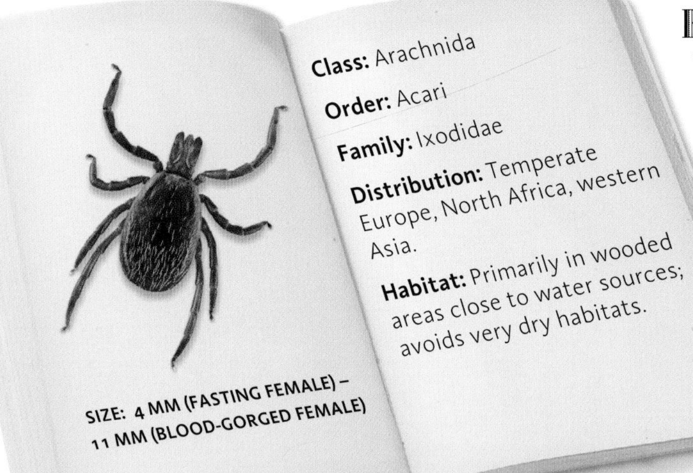

Reproduction and distribution

Ticks practice an indirect form of reproduction, as males do not have sex organs. After "preparing" the female's genital opening with his mouthparts, the male deposits a spermatophore containing its semen. The fertilized and blood-gorged female then deposits its clutch of eggs on the ground. The blood is necessary for egg maturation.

Rural vampires

The development of *Ixodes* consists of three stages – larva, nymph and adult. In order to transition from one stage to the next, the tick must have a blood meal, let itself fall to the ground, molt and then find another host to feed on. Since the waiting period between hosts can be quite long, the full life cycle generally lasts a minimum of two to three years. Since this arthropod is very resilient it can remain a juvenile for several months – even several years.

A tick lies in waiting on top of a blade of grass with its front legs extended. These legs are equipped with sense organs that detect carbon dioxide and the heat emitted by its victim. The tick pierces the skin of its host (bird, reptile or mammal – preferring large adult animals) to suck blood. The larval blood meal can last two to three days, and that of an adult female up to eight days. Males do not nourish themselves; they live on reserves accumulated as nymphs.

Origin of the name

Also known as the Sheep Tick. The ancient Egyptians called ticks *ricinus* due to the resemblance of blood-gorged females to castor bean seeds (castor beans belong to the genus *Ricinus*). This name was adopted by several Greek authors, including Aristotle, and has persisted through the centuries. Linnaeus classified the common tick *Acarus* (from the order *Acari*) *ricinus*, until Latreille gave it its present, very evocative name: *Ixodes* ("sticky" or, by extension, "tenacious, attached"). The word "tick" is of Germanic origin.

A giant among dwarves

Ticks are acarians, like harvest mites of the genus *Trombicula* (those minuscule red "spiders" that occasionally attack green plants). They have four sets of legs, like scorpions and spiders, placing them in the class of Arachnids. Adult ticks also have eight feet but the larva only has six. Among the Acari, however, which are generally very small, they are giants. *Ixodes ricinus* is the most widespread tick in Europe.

WARNING: DANGER!

Tick bites are not without risk, since these insects carry a number of infectious germs. In many countries in the Northern Hemisphere, the most common is the bacteria responsible for borrelliosis, or Lyme disease. It is important to immediately remove the tick with pincers or forceps, then apply disinfectant. Other product applications, etherizing the tick, or rubbing it with skin cream, are not recommended since that can lead to additional release of germ-ridden saliva. If a red rash develops around the bite, you must see a doctor as soon as possible.

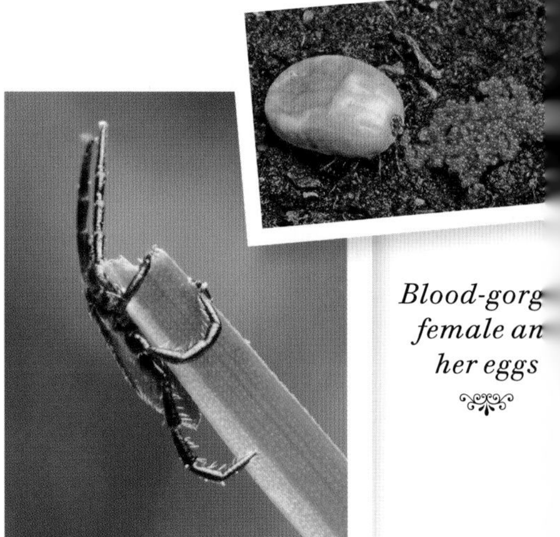

Blood-gorg female an her eggs

Waiting to attack a host

Ixodes ricinus

(Linnaeus, 1758)

Size: 4 mm (fasting female)–
11 mm (blood-gorged female)

Lampyris noctiluca

Glow-worm

Glow-worm larva

Class: Insecta

Order: Coleoptera

Family: Lampyridae

Distribution: Temperate Europe and Asia.

Habitat: Edge of wooded areas, humid prairies, embankments, bushes and gardens.

SIZE: MALE (LEFT): 10–12 MM
FEMALE (RIGHT): 15–18 MM

Luminous loves

During warm summer nights in June and July, the female Glow-worm climbs in grasses. At the top of a blade of grass, she bends her abdomen and extends it as far as possible to expose her photophore (photo at bottom right). Should a male happen to be flying by and see her luminous signal – thanks to his oversized globular eyes – he will land next to her lovely glowing form. After mating, the female deposits a hundred or so lightly luminescent eggs (about 1 mm in diameter) on the surrounding ground and vegetation.

Snail predators

The eggs hatch in about 30 days. The larva is a voracious predator of snails. After inserting its head into the shell, the larva injects an anesthetic into its prey and then releases digestive enzymes from its saliva. All that is required after that is to digest the snail's liquefied remains. For reasons that are not clear, the larva may also emit light signals (though much fainter that those of the adult female). Having already undergone two molts by the onset of winter, the larva then burrows underground. It will take two more molts and one more season of feeding on mollusk prey before the larva will hibernate for a second time. By springtime of the second year, the larva is about 3 mm in length. It has transformed into a nymph, at which point its elytra reveal that it is a male. The adult female still looks like a larva, though she lacks the clear patches at the tips of her abdominal segments.

Origin of the name

Lampyris (from the Greek *lamp*, for "glowing," and *pyros*, for "fire") was designated the Glow-worm by the Romans. The species name, *noctiluca*, means "which glows in the night" in Latin. Though clearly not a worm, the female is wingless and therefore resembles a larva. The male, on the other hand, has typical coleopteran elytra.

Cold light

Female glow-worms have a photophore in their abdomen – a light-emitting organ that produces an intense yellow-green light. This phenomenon, known as bioluminescence, is due to the oxidation by air of luciferin molecules under the control of the enzyme luciferase. This reaction generates carbon dioxide and releases energy in the form of light. The Glow-worm controls its photophore by regulating the flow of oxygen. This is known as "cold light" since all the energy produced is released as light, rather than heat. In contrast, a classic incandescent light bulb releases 95 percent of its energy as heat and only 5 percent is converted into light.

A shortage of light in the countryside

Glow-worms have been in sharp decline for several decades. This great mollusk predator should be appreciated and protected by gardeners. Unfortunately, that is not the case. Glow-worms are often the victims of insecticides and lawn mowers. Air pollution also plays a significant role in their decline. It is sad that only the older generations recall the times when our paths were illuminated by these enchanting little lights ...

"Glowing" female

Egg-laying

Lampyris noctiluca

(Linnaeus, 1767)

Size: Male (right): 10–12 mm

Female (left): 15–18 mm

ACTUAL SIZE

LEPTINOTARSA DECEMLINEATA

Colorado Potato Beetle

Potato beetle larvae

Highly prolific

After hibernating underground at a depth of 10 to 20 inches (25 to 50 cm), adults emerge at the beginning of spring and mate. The females lay groups of 20 to 30 yellow-orange eggs on the underside of leaves. The larvae quickly start feeding voraciously on the foliage. After their third molt, the larvae leave the plant, drop to the ground and bury themselves in the soil to metamorphose into nymphs, and then into adults. Often there are as many as two generations per year. Females can live up to two years and produce more than 2,000 eggs. It is easy to see the resulting proliferation.

Chronicle of an invasion

The first mention of the Colorado Potato Beetle was made by Say, who described it in 1824. The entomologist observed it in the foothills of the Rocky Mountains in the United States, where it was feeding discretely on wild Solanaceae plants. The precise geographic origin of this beetle remains uncertain. It is quite possible that it originated in more southern areas (perhaps Mexico) and was brought northward by the Spaniards. The phenomenal expansion of the Colorado Potato Beetle can be attributed to cultivation of potatoes by the original colonists. The insect reached the Atlantic coast in 1874 and appeared in Europe in 1877. Several additional, more or less successful invasions, followed until the one of 1922, at the port of Bordeaux, which signaled its definitive final entry into Europe. At present, this beetle is found pretty much everywhere in the world where potatoes are grown.

Mating

Origin of the name

This insect was originally placed in the genus *Doryphore* (literally meaning "lance carrier"), a term used in antiquity for soldiers carrying a spear. This name was used entomologically to designate certain North American coleopterans covered with long points. Over time, however, this species was renamed *Leptinotarsa* (or "thin muscled tarsus" in Greek) in reference to its feet, and *decemlineata* (from the Latin for "10 lines") in reference to the 10 dark bands on its elytra.

A big potato eater

The Colorado Potato Beetle is closely related to the Rosemary Beetle (p. 80). It grows on various plants of the Solanaceae family (mostly on potatoes, but also climbing night-shade, deadly night-shade, tobacco, tomatoes, eggplant, and so on). The Potato Beetle larva is not affected by the toxic alkaloids found in the leaves and stems of these plants. Only *Datura* plants occasionally attract this insect and poisons its larva.

NEVER LET YOUR GUARD DOWN

Despite massive use of insecticides, and the resulting collateral damage it has done, resistance to pesticides has developed in these insects. Still, a gradual decline in Potato Beetles has occurred, to the point where many young gardeners have never seen it. It is important to remain vigilant, however. Should the pest reappear, it must be mechanically destroyed as quickly as possible.

Laying of eggs

Leptinotarsa decemlineata

(Say, 1824)

Size: 8–12 mm

LIBELLULA DEPRESSA

European Dragonfly

Libellula larva

Class: Insecta

Order: Odonata

Family: Libellulidae

Distribution: Europe and Western Asia.

Habitat: Small stagnating pools of water (marshes, ponds and lakeshores).

WINGSPAN: 75 MM
SIZE: 39–48 MM

ORIGIN OF THE NAME

The name *Libellula* is the diminutive form of the Latin *libella*, designating an architect's level. Inspired by Guillaumme Rondelet, the naturalist, Linnaeus called the dragonfly larva *Libella fluviatilis* in 1554, due to its similarity to the marine fish *Zigaena* (the hammerhead shark), whose shape resembles the level used by architects. The species name, *depressa*, used without any psychological connotations, refers to the insect's large, flat abdomen.

A TRUE LIBELLULIDAE

The European Dragonfly is a true Libellulidae (in the suborder Anisoptera), and not a damselfly like the Damselfly (p. 58). When at rest, the adult's wings are extended from both sides of the abdomen. The stocky and burrowing larvae breathe and propel themselves thanks to designated tracheal gills not found in damselflies. The abdomen of the European Dragonfly is yellow at the beginning in both sexes, but then takes on a bluish tinge in adult males with yellowish lateral crescents.

POND HUNTER

The European Dragonfly likes small pools of stagnant water, including garden puddles and large marshes. It also frequents ponds, lakes and dead-end branches of rivers. Adults live near the water's edge, often on top of branches. They can also be found in forest clearings or on hedges, hovering before rapidly diving in pursuit of small insects (mosquitoes, wasps and butterflies). Males jealously defend their territory against rivals. After mating, the female lays her eggs on floating aquatic vegetation. Her technique is very specific: she hovers just above a leaf, and deposits one of her eggs by briskly bending her abdomen down each time.

CARNIVOROUS LARVAE

The larvae emerge after four or five weeks. They hunt lying in wait in the mud. When potential prey passes in front of it, the larva first extends its "mask" (a lower lip covered with hooks) in order to grasp it. In cases where the pond has dried out, the larva buries itself in the mud and enters a state of lethargy. It can survive this way for more than a month. It spends the winter in diapause deep in the mud. The following spring, having reached full growth, the larva waits for a sunny day to metamorphose. After pulling itself out of the water, it holds onto nearby vegetation and emerges as a young adult after its skin has dried. Still pale, the young adult deploys its wings through "blood" (hemolymph) pressure and waits for them to harden in the sun. It takes about 10 days for it to attain sexual maturity.

FLYING MONSTERS

Along with the Ephemeroptera, dragonflies are considered among the oldest winged insects. They first appeared more than 300 million years ago, during the carboniferous period (before the dinosaurs). The atmosphere in the immense forests of that time was much richer in oxygen than today. This allowed dragonflies to attain considerable sizes – more than 28 inches (70 cm) in length!

The female is yellowish in color

ACTUAL SIZE
N. Libellula depressa
Linnaeus, 1758
Wingspan: 75 mm
Size: 39–48 mm

Lilioceris lilii

Scarlet Lily Beetle

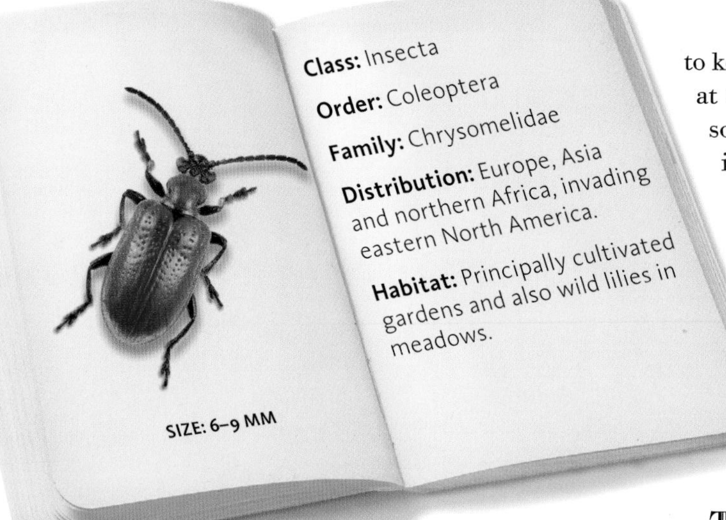

Leaf beetle nymph in its protective capsule

Origin of the name

This insect was originally called *crioceris*, which means "ram horn" – no doubt in reference to its ringed antennae. These beetles live on plants of the Liliaceae family and were subsequently reclassified in a separate genus: *Lilioceris*. This name is a contraction of *lilium* ("lily" in Latin) and *crioceris*.

Attacking lilies

Among the Coleoptera, members of the Chrysomelidae family often evoke contradictory sentiments. On one had we admire the beauty of their colors (see, for example, the Rosemary Leaf Beetle, p. 80), but we dread the damage they can do (best exemplified by the Colorado Potato Beetle, p. 154). Certainly the Scarlet Lily Beetle is never appreciated by gardeners, since this bug feeds exclusively on Liliaceae plants, with a strong preference for lilies. These include those grown in our gardens, in addition to the rare and protected mountain species *Lilium matagon*, but disregards the Lily of the Valley.

Lilies under surveillance

In order to avoid beetle infestations and the need for insecticides, it is necessary to keep close watch on lily fields at the beginning of spring. As soon as adult beetles appear it is best to suppress them before they reproduce. If they have started laying eggs already, both eggs and larvae have to be destroyed. Even though its vermillion red coat contrast with the vegetation, it is not easy to surprise these bugs since at the slightest sign of danger they hide underneath the leaves. If you try to catch them by hand, they quickly drop to the ground before you have a chance to grab them.

Eggs and larva

There is no time to lose

From the first pleasant days of spring, the adults having passed the winter underground, they emerge and look for lilies in order to mate. Gravid females deposit some 300 bright red eggs in small bunches held in place by a viscous material under the leaves. The reddish larvae that emerge from there are voracious and perform no other function than to devour the leaves of their host plants. Thanks to an anal orifice on their backs, the larvae are covered by their own excrement, which keeps them cool and protects them against predators. This novel type of behavior is also found in other Chrysomelidae, like the Thistle Tortoise Beetle (p. 66). Since the plants become etiolated, the larvae complete their development in only three weeks. They subsequently drop to the ground and bury themselves, perfectly invisible in their white silk cocoons covered with earth. The developmental cycle of these beetles is fast enough for them to produce three generations per year.

The Scarlet Lily Beetle's Complaint

This beetle is exclusively vegetarian, so you can seize it without fear. Should you do this, hold it between your fingers without crushing it and bring it near your ears. You will be able to hear its strange "cry." When this bug is disturbed, it emits a unique sound created by a chirring organ located between its abdomen and elytra.

Lilioceris lilii

(Scopoli, 1763)

Size: 6–9 mm

Lithobius forficatus

Brown Centipede

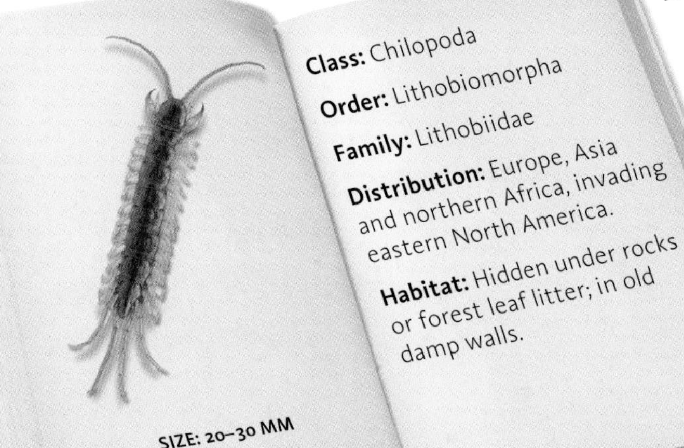

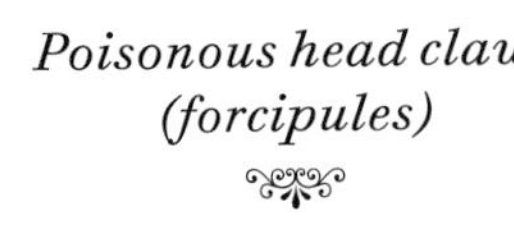
Poisonous head claw (forcipules)

Origin of the name

The Lithobiidae get their name from their lifestyle: in Greek, *lithos* means "stone" and *bios* means "life." These are basically species that live under rocks. Its species name comes from the Latin *forfex*, meaning "pincers," referring to the impressive pair of forcipules, or claws, common to all chilopods.

An agile centipede

The Brown Centipede belongs to the class Chilopoda, which also includes the House Centipede (p. 222). Although the names millipede and centipede are often interchanged, the term centipede (or "hundred footed") for these animals, is a more correct term since they really only have a few dozen feet. These creatures are photophobic and like humidity. We often find them under rocks or a big piece of fallen wood fallen. Despite poor vision, they move with speed and agility thanks to their antennae. Prolific hunters, centipedes capture their prey alive and pierce it with their forcipules prior to injecting it with venom and devouring it.

The season for love

Reproduction begins toward the end of winter. Pairs seek each other, move apart and then rejoin, all while palpating each other with their antennae. The length of this courtship attests to the attraction the two partners have toward each other. The couple eventually retires to a hiding place to mate. Centipedes don't really mate since they practice an indirect method of fertilization. Semen is produced in the male by two small glands near its genitals. The male deposits a tiny drop of semen onto a small layer of previously woven silk called the spermatophore. The female seizes this and brings it to her genital opening with two small hooks known as gonopods. Several dozen eggs are subsequently deposited. Before leaving the eggs, the female uses its gonopods to carry them one by one to a "nursery" and then wraps them in a small cocoon of earth. After emerging, the young centipede has only seven pairs of legs. It's only after about 10 molts that it becomes an adult with 15 pairs of legs. Unless eaten by predators, a centipede can expect to live five years, during which it undergoes several additional molts.

Not all appendages are the same

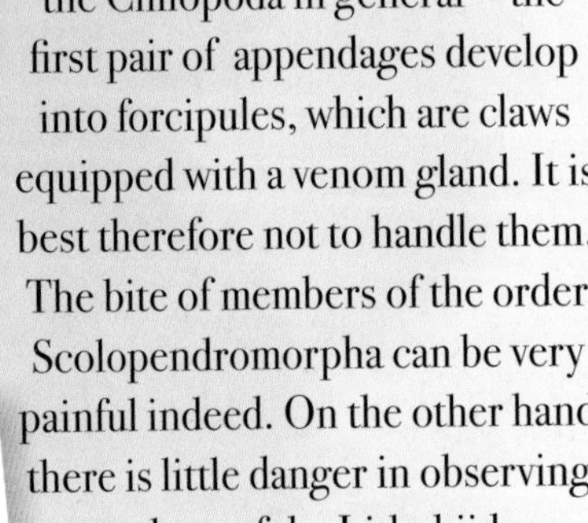
Among the Lithobiidae – and the Chilopoda in general – the first pair of appendages develop into forcipules, which are claws equipped with a venom gland. It is best therefore not to handle them. The bite of members of the order Scolopendromorpha can be very painful indeed. On the other hand there is little danger in observing members of the Lithobiidae, as they try to flee immediately and don't seem to be aggressive biters.

ACTUAL SIZE

Lithobius forficatus

(Linnaeus, 1758)

Size: 20–30 mm

LUCANUS CERVUS

Stag Beetle

The Stag Be
"doe"

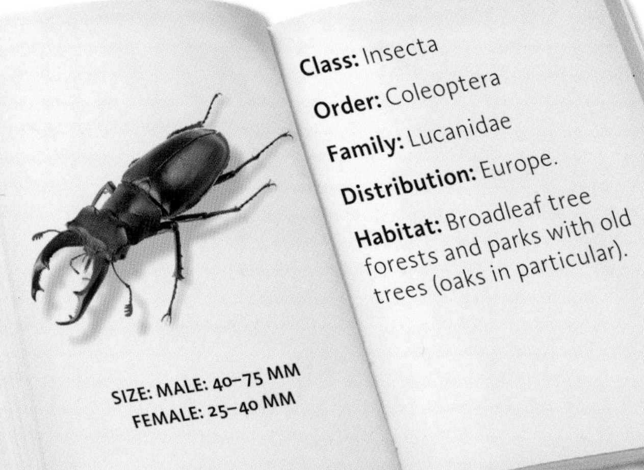

THE TITAN OF BROADLEAF TREE FORESTS

The male Stag Beetle is the largest insect in Europe. When threatened, it hoists its enormous mandibles in a menacing manner. However, the male is completely harmless as it cannot close or tighten its mandibles. Beware, though, of the females (or "does") – they have poorly developed but much more powerful mandibles!

The Stag Beetle feeds on tree sap or on very mature fruit. Its variable size depends on the quality of nutrition that the larva had during its development. It can survive several weeks without eating, dipping into the reserves it gathered during its larval existence. It is a crepuscular insect that loves broadleaf tree forests. In June to July, often in stormy weather, it's possible to observe the male's flight, conspicuous by its impressive "motorlike" sound and its typical pattern: raised elytra to allow the beating of its wings, mandibles pointed skyward to keep its equilibrium and legs splayed, ready for landing.

A Stag Beetle larva

ORIGIN OF THE NAME

Lucanus comes from the Latin *lucus*, meaning"sacred woods" – a reference to this insect's behavior. The male's superb, ramified mandibles remind us of a stag's antlers, the fact that gave it its species name of *cervus*. They are also given a common name of kite (in French) because, in times gone by, it was thought that children used to tie a string to one of the beetle's legs and let it fly like a kite.

HIDDEN LARVA

Stag Beetles emerge and reproduce in springtime. After mating, the female lays about 20 eggs under the rotting stump of an old tree (oak, willow, chestnut, cavity in a garden fruit tree, and so on) and dies soon after. After hatching, the larva lives in hiding for a span of three to five years, during which it feeds on decomposing dead wood. Metamorphosis occurs in the fall, either on site or underground, in a shell constructed by the larva.

Nymphs. The male (below) already has elongated mandibles

Two males in combat

A SPECTACULAR FIGHTER

Just as stags confront one another during the rutting at the beginning of summer, male Stag Beetles with large "racks" fight at dusk, during warm evenings. Using their oversized mandibles, they fight to stave off their rivals and flip them over on their backs. Since females are three times less numerous than males, the latter must put up a good fight to win the favor of a "doe." Males with puny mandibles (the *minor* form) therefore have very little chance of mating ...

LEGENDS AND REALITY

In days gone by, people used to attach Stag Beetle mandibles around children's necks as either good luck charms or to ward off illnesses. It was also believed these insects had the power to transport glowing embers in their mandibles and thus set fire to houses. This superstition was once very widespread (in some years veritable "raining of stag beetles" was observed!) but is now rare due to the disappearance of favorable habitats. Today, the Stag Beetle is a protected species.

ACTUAL SIZE

Lucanus cervus

(Linnaeus, 1758)

Size: Male: 40–75 mm

Female: 25–40 mm

Green Bottle Fly

Juvenile, without melanin

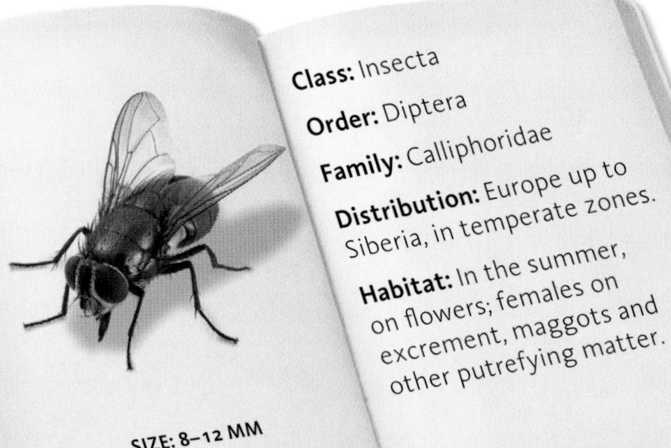

The male Green Bottle Fly (eyes very close together)

Origin of the name

The genus name *Lucilia* is derived from the Latin *lux* or *lucis*, meaning "light," in reference to the brilliant colors of these flies. The species name, *caesar*, means "imperial" (the title given to Roman emperors). This species is often called the Golden Fly or the Green Fly, although the true Green Fly (or Meat Fly) is another species altogether (*Lucilia sericata*).

On flowers and cadavers

At the first signs of spring, the Green Bottle Fly starts warming itself on sunlit walls. This pretty metallic green insect takes on a coppery tone with age. It is fairly widespread, although most commonly found near housing. It is attracted by the scent of flowers and other strong-smelling objects, including various mushrooms with large caps (whose spores the fly helps disseminate) and decaying animal cadavers that its strong sense of smell discovers. The females come by to mate and lay their eggs. The resulting maggots (a name reserved for the legless larvae of the order Diptera) feed on decomposing carcasses. This is a very prolific species, since a single female can have successive egg-layings that may number in the thousands! The majority of these, however, end up devoured or else the victims of bad weather. The life cycle of this fly is quite short, lasting only a few weeks. Larvae regurgitate digestive juices and then "drink" the liquefied cadaver tissues. During metamorphosis, the larva buries itself in the ground and becomes immobile. Its skin will then harden and the larva transforms itself into a pupa that protects the nymph. This process is similar to what occurs in the common Domestic Fly (p. 178). Upon emerging, the young fly does not yet have the brilliant metallic coloration.

Medical and legal experts

The Green Bottle Fly is one of the first insects to appear and to lay its eggs in decomposing cadavers, which explains its importance to forensic science investigations. Since these insects succeed one another in a very ordered and predetermined manner, the presence of one or another species of larvae allows medical examiners to determine the time of death. A majority of these flies belong to the first "squad" – they appear very quickly after death, even before the cadaver emits any odors whatsoever. *Lucilia caesar*, however, forms the second wave, appearing when the cadaver starts emitting odors. Additional "squads" of flies will succeed one another during the three first months that follow death. While they play an active role in recycling decomposing matter, these insects also play a significant role in criminal science investigations!

MAGGOT THERAPY

From time immemorial it has been known that the Green Fly (*Lucilia sericata*) feeds only on decomposing meat, but not living flesh. Today, these maggots are used to clean the wounds of patients. These flies devour the dead tissues, and their secretions have been shown to have excellent scarring and antibiotic properties. The sterilized maggots are placed on the wound and covered with gauze. All that is left is for the patient is to overcome the emotional aspect of this treatment. Once this is conquered, we can predict that maggot therapy may have a promising future.

Maggots and pupae of Green Fly

ACTUAL SIZE

N.°

Lucilia caesar

(Linnaeus, 1758)

Size: 8–12 mm

Macroglossum stellatarum

Hummingbird-hawk Moth

Moth caterpillar with a posterior "horn"

Origin of the name

Here is a moth with no shortage of common names! In French, *Moro-sphinx* means "crazy sphinx," referring to its extremely fast flying ability. It is also called *macroglosse du caille-lait*, with reference to its primary host plant, Yellow Bedstraw (*Galium verum*). This moth's appearance, color and characteristic flight has also inspired several bird names, including the Sparrow Sphinx, Bird Beak, Sphinx Colibri, Sparrow Hawk and, of course, Hummingbird-hawk. The genus name *Macroglossum* (from the Greek *macro*, or "large" and *glossa*, or "tongue") refers to the moth's long proboscis; and the species name refers to *Stellaria*, the plants that the caterpillar feeds on.

A familiar migrator

Spread from southern Europe and North Africa to central Asia and Japan, this small moth is indigenous to the southern regions of its domain but migrates further north. Each summer it reappears in our region very early in spring and then delays departure later in the season. In addition to its ability to fly in place (see below), the Hummingbird-hawk Moth can fly in a straight line at great speed, allowing it to travel long distances. In southern France, two generations per year are possible, while in the north new migrants are required because none survive the winter.

A daylight night flier

The Hummingbird-hawk Moth can often be seen flying in gardens and hedges in full sunlight. They fly mainly in the afternoon, but never at night (which is unusual for moths, since most species are nocturnal). When night comes they hide in cracks in rocks or in houses, where they can not only rest unnoticed, but their dull, shaded forewings can cover the orange hindwings. When it is still cool in the morning, they keep warm by vibrating their wings before leaving to gather nectar.

This moth does everything while flying

After mating (in flight!) and a period of feeding, the female looks for groups of Yellow Bedstraw plants in bright sunlight and deposits about 200 small green eggs. Still flying, she inspects each plant one by one and deposits one egg into each flower bud; the eggs themselves resemble an unopened bud.

The caterpillar, either brown or green in color, has a small posterior horn typical of most Sphingidae. While they feed primarily on bedstraw, they are also found on other *Rubiaceae* plants and *Centranthus*, *Stellaria* and *Epilobium*.

The Hummingbird of Insects

It is difficult to obtain clear photographs of this moth since is always in motion. While hovering and moving rapidly from flower to flower, it gathers nectar at a frantic pace. This type of hovering like a hummingbird is accomplished through extremely rapid wing beats – 75 per second! With its very long 1-inch (2.5 cm) proboscis, moreover, it gathers nectar from flowers with deep corollas, including jasmine, *Buddleja*, violets, petunias, lilacs and so on.

Macroglossum stellatarum

(Linnaeus, 1758)

Wingspan: 40–55 mm

ACTUAL SIZE

MACULINEA ALCON

Alcon Large Azure Blue Butterfly

Class: Insecta

Order: Lepidoptera

Family: Lycaenidae

Distribution: Europe — very localized due to its ecological niche.

Habitat: In heaths, swamps and peat bogs.

WINGSPAN: 32–40 MM

ORIGIN OF THE NAME

The word "azure" applies to several species of Lycaenidae with blue wings. The scientific name, a contraction of the Latin *macula*, for "spot," and *linea*, for "line," refers to the genus-specific markings on the upper surface of this butterfly's wings. Many butterflies have been assigned a species name based on a mythological character. That is also the case for the name Alcon, who was thought to be a famous Cretan archer, father of the Argonaut Phaleros.

A DEMANDING PLANT AMATEUR

The Alcon Large Azure Blue Butterfly has an astonishing life cycle. Its existence depends on the simultaneous presence of a unique species of plant and genus of ants! Without these two "partners," this butterfly has no chance of reproducing. First, the Alcon Large Azure Blue Butterfly must find a budding flower to lay its eggs. This can't be just any kind of flower but, specifically, the Marsh Gentian found in moors and marsh-lands – one of the few gentians that grow in plains. The female lays her pure white eggs (smaller photo to the right) on the flowering bud. Upon hatching, the the minuscule caterpillar penetrates the flower's corolla, where its small mouth allows it to feed exclusively on the flower's ovaries and stamens.

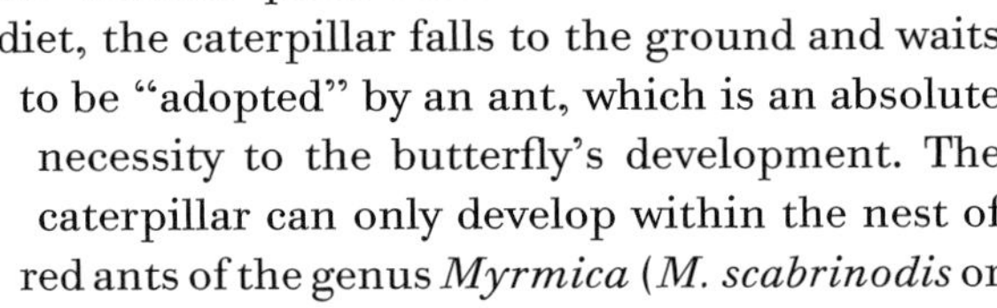

An Alcon Blue Butterfly posed on a Marsh Gentian flower

ANTS AS FOSTER PARENTS

After three to four weeks of this delicate plant-based diet, the caterpillar falls to the ground and waits to be "adopted" by an ant, which is an absolute necessity to the butterfly's development. The caterpillar can only develop within the nest of red ants of the genus *Myrmica* (*M. scabrinodis* or *M. ruginodis*). To convince its potential future family to adopt it, the caterpillar must first make a solid case for it. In addition to mimick-ing a larva, it also secretes pheromones (allomones) that closely match those of ant larvae and thereby fools the fearsome female worker. Having accomplished this and both convinced and stimulated them, the female ant workers feed the "interloper" as if it was one of their own! Pampered throughout the winter, the caterpillar grows and transforms itself into a chrysalis by springtime. Dur-ing the summer, the chrysalis becomes a butterfly and leaves the ants' nest. The male and female butterflies devote their short (one month) lifespan to reproducing, in a time frame that coincides with the period during which the host plant is full of buds.

THE THREATENED AZURE BUTTERFLY

The very fact that the peculiar life cycle of these butterflies requires both a plant and an ant host makes them quite vulnerable. Also drought, and the destruction or disappearance of habitats have brought about a drastic reduction in the two host species that these butterflies depend on for their very existence. For a few decades now they have been disappearing in many localities, although in France they remain widespread but not very abundant. Several efforts at protecting this butterfly are underway.

The larva and eggs of the Azure Butterfly

Maculinea
alcon
(Denis &
Schiffermüller,
1775)
Wingspan:
32–40 mm
ACTUAL SIZE

Praying Mantis

Juvenile mantis

SIZE: MALE: 42–61 MM
FEMALE: 48–75 MM

Class: Insecta

Order: Mantopterae

Family: Mantidae

Distribution: Europe except for the north; North Africa and to the south of the Sahara; temperate Asia; introduced to the United States.

Habitat: Dry and warm areas with vegetation; scrublands, moors, gardens, wastelands.

Origin of the name

The Praying Mantis derives its name from its obvious posture. In Latin, Mantis means "soothsayer" and *religiosa* is linked to the position of the forelegs, which mimic a person in prayer. This point has touched human thinking over the years and in many countries: for example, the expression "Prie-Dieu" is used in Provence as a common name for this insect; in Creole, it is called a "Cheval-Bondié"; in Spanish, "Saint Theresa beast"; and in English, the "soothsayer."

Fast growth

Praying Mantises go from egg to adult in the short space of three months – a very rapid pace for such a large insect. The larvae are born in May to the beginning of June and crawl all the way to the opening of the ootheca, whereupon they immediately molt into miniature replicas of the adults, though without wings. They remain gregarious in behavior during the first few days and then become solitary.

Juvenile mantises (see photo above), carnivorous like their adult counterparts, live in low-lying bushes or on the ground and go through seven to eight molts, the last one of which occurs at the end of August. The males live for one or two months and risk their lives every time they mate. Females live a bit longer, for three to four months, and are completely devoted to the laying of their numerous eggs.

Once she has collected sufficient reserves, the female goes about finding a spot – under a sun-baked rock, a heather branch or the underside of an abandoned shutter – to build her ootheca. This "egg mass" is very well built. At its center there are 150 to 300 eggs, in ordered rows opening up to the exit, followed by a series of thin strips covering each other like flooring tiles. Surrounding all this is a thick layer of solidified foam that protects the egg mass from the cold and from drying out. This process will be repeated by the female two to three times before the end of November.

Ferocious huntress

The Praying Mantis is a fearsome predator. The majority of its prey is made up of insects, often quite mobile and bigger than it is. The mantis excels at ambushing as well as chasing, or in taking "chameleon-like" steps.

The Praying Mantis gets the upper hand on its prey with formidable anterior legs well equipped to seize and grasp any insect. She holds onto the prey with the spiny tibia and femur sections of her legs, which fit together perfectly much like the mouth of a crocodile (photo below).

Babies emerging from the ootheca

A FABULOUS INSECT

Following a good meal, the Praying Mantis uses her anterior legs, equipped with a small brush of transparent hairs, to completely clean its head, exactly like a cat would do with its paws! This unusual insect has given rise to many popular superstitions. It could lead the way, some believe, for children lost in the forest! Certainly its very mobile neck and its enigmatic stare leave a strong impression!

A grasshopper on the menu

ACTUAL SIZE
Mantis religiosa
(Linnaeus, 1758)
Size:
Male: 42–61 mm
Female: 48–75 mm

Violet Blister Beetle

Larvae on a flower

ORIGIN OF THE NAME

If we ignore the origin of the name *Meloe*, coined by Linnaeus, we know that in ancient times this insect was called *bupreste*, from the Greek, meaning "which bloat cattle." The mistaken implication was that it caused often fatal swelling in cattle. In the Middle Ages it was also called "cattle bloater." This probably stems from confusion with insects of the Cantharidae family, which can reportedly cause such symptoms.

CONGENITAL OBESITY

Thanks to its striking appearance it is impossible to confuse this beetle with any other; its bulging elytra are so short that they only cover a small portion of its rounded abdomen (particularly among females). This wingless insect is found pretty much everywhere, though not common, both in and near hedges, on flowers, particularly buttercups. After a brief appearance in May, it disappears in the month of June. Its highly varied size is linked to the quality of nutrition the larvae obtain. The male, distinguished by its angled and knobby antennae, is smaller than the female, whose abdomen is dilated by the large number of eggs she carries.

BEE PARASITES

Blister beetles have a complex and original life cycle. In the spring they mate extensively on flowers. Afterward, the female digs a small cavity in the soil and deposits several thousand eggs inside it. These hatch about one month later. The minuscule larva is very mobile and called "triungulin" because of three flat markings on each foot. The larva emerges from the ground, climbs atop a flower and awaits the arrival of a solitary bee that it attaches itself to (see photo at bottom left). When returning to the hive, the bee carries its clandestine passenger (sometimes called "bee flea") with it. The larva then leaves its carrier and becomes enclosed in a cell containing a bee egg. After devouring the egg, the larva transforms itself into a legless form resembling a white worm and gorges itself on honey to the point of obesity. It spends the winter immobilized as a pseudo-nymph. The following spring, the pseudo-nymph gives way to another larva that rapidly changes into a nymph and then an adult.

An encounter between a larva and a bee is hazardous, and most triungulins attach themselves to other species feeding on flower nectar. By depositing a very large number of eggs (sometimes as many as 4,000!), the female blister beetle increases her chances of reproducing.

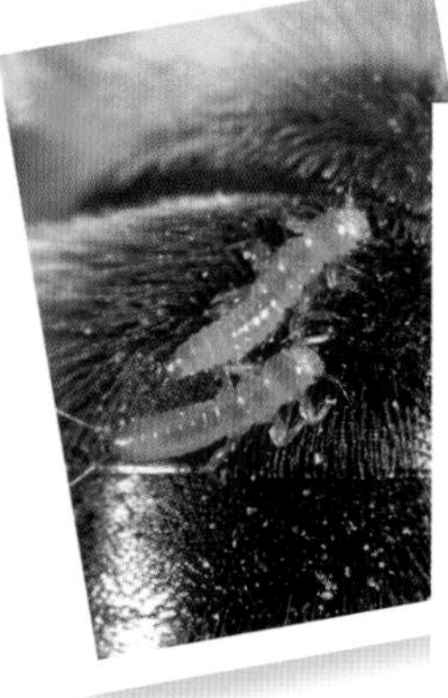
Bee carrying two larvae

Mating

PLUMP ... BUT INEDIBLE

With its corpulent body and lack of wings, the blister beetle appears pretty vulnerable. Nothing could be further from the truth...When it senses itself in danger, it releases a yellowish liquid that stains and leaves a tenacious odor on your fingers. It must have a foul taste as well because it turns birds away. The Spanish fly (*Lytta vesicatoria*), a related species, produces cantharidin, which is much more toxic, and whose blister-causing properties (and alleged aphrodisiac) have been known since antiquity.

ACTUAL SIZE
Meloe violaceus
Marsham, 1802
Size: 11–35 mm

Melolontha melolontha

European Cockchafer

Female Cockc

Class: Insecta

Order: Coleoptera

Family: Scarabaeidae

Distribution: Europe except for the Iberian Peninsula and the northern part of Scandinavia.

Habitat: Dry and warm areas with vegetation; scrublands, moors, gardens and wastelands.

SIZE: 20–30 MM

Cockchafer Larva

Origin of the name

This insect is also known as a Dung Beetle. *Melolontha* is originally the Greek name for a certain insect (probably a *Cetonia*) that children played with and that frequented the orchards of *mélon*, the apple tree. The species and genus names are identical. The first people to identify them in the 18th and 19th centuries often used repetition (tautology) to name the most numerous species of a given genus (an example: *Milvus milvus*, the Red Kite, a bird of prey).

Cyclical sightings

Today, the "common" European Cockchafer is either not very abundant or even rare locally. However, up till the 1960s (before the systematic use of pesticides), massive sightings of this insect every three (or four years more to the north) were greatly feared. The larvae caused huge amounts of destruction in farm fields of cereals, beets, strawberries and especially of pine tree farms.

Following hibernation, adults emerge from the ground at nightfall at the end of April to beginning of May and feed on leaves for two to three weeks before mating. Egg-laying (20 to 30 eggs) occurs in the ground. There are three larval stages, each following after a molt. Every winter, the larva digs itself deeper and deeper, and comes up again at springtime. Having finished its development in the second year of its life cycle, the larva will metamorphose in a little shell of soil, sealed with saliva. The still soft adult spends the winter underground in diapauses and will only emerge in the spring. This long multiyear development plan explains the cyclical appearance of this species – the infamous "cockchafer years."

Undemanding white worms

Equipped with strong mandibles, the Cockchafer larvae devour the roots of a large number of plants. Unlike *Cetonia* larvae, this larva is undesirable; therefore it becomes important to distinguish between these two (see p. 74 for the Golden Cetonia). Adult Cockchafers shy away from excessive heat. Therefore, we can see more of them at sunset and hear them humming around trees, or under streetlights since they are keenly attracted to light.

Not taken by the cockchafers

The ever presence of this insect in times gone by gave rise to several popular expressions relating to it. For example, "not bitten by cockchafers" is said of something quite unique and remarkable in itself; someone might also be described as "dizzy as a cockchafer" since these insects often bang into obstacles and fall loudly to the ground.

Cockchafer Proliferation

In 1574 in England, cockchafers were so numerous it was said they stopped windmills from working. The chubby larvae were once called "chicken fatteners," since chickens and other fowl were brought out to fields after harvesting to eat the "uprooted" larvae. During the 19th century, the French government encouraged "cockchafering," which consisted of shaking tree branches during daytime so the cockchafers fell off and could be killed. In 1835, in the Sarthe region, 300 million insects were captured using this method! In 1887, it was proposed to add incentives. In the Mayenne region, for example, officials offered 2.5 francs per quart of killed cockchafers.

The takeoff

ACTUAL SIZE

Melolontha melolontha

(Fabricius, 1775)

Size: 20–30 mm

Misumena vatia

Goldenrod Crab Spider

Another Thomisidae
Misumenops tricuspid

Origin of the name

In Greek, *Misumena* signifies "detestable" – not a very nice depiction by Latreille, who first described this spider. The species name *vatia*, "with bowed legs" in Latin, refers to the outstretched position of its legs when it is hunting.

These arachnids are presently called "crab spiders" because of their crablike walk. They family name Thomisidae (from the Greek meaning "linked by a cord") was given because it was thought that they wrapped their prey in silk threads like many other spiders do. Not so, however – they don't do this. They were assigned this name erroneously before their actual behavior was understood.

Deadly venom... for insects

The female Goldenrod Crab Spider hunts stealthily, hidden within a flower. Thanks to her coloration, she is invisible amid the flower petals and waits, ready to pounce on an insect victim. She does not avoid sizeable prey – big flies, moths, Syrphidae and bees. Seeking nectar, potential victims approach the flower and do not see the danger that lurks there. Seemingly out of nowhere, the spider seizes and bites its prey in the nape of its neck, and the venom delivered near the head ganglia causes rapid paralysis. This also serves to prevent any possible counterattack by the prey insect that, while attempting to flee, could take the spider with it, or else sting it, depending on the species involved.

Mistress of camouflage

In order to render herself invisible to her prey and predators, the female crab spider can adapt her coloration to match that of any white or yellow flower in which she resides! This adaptation, probably linked to vision, is accomplished through the transfer of a pigmented fluid in the cuticle of her abdomen. If the flower is white, the spider excretes the yellow pigments in its cuticle so as to become completely white in about three to six days. If she moves onto a yellow flower, it takes about 10 days for her to become yellow again since she has to synthesize new pigments, which takes more time.

A fly caught in the trap

The male

Small wondering male

The male crab spider is a very discreet flower visitor. Darker in color and considerably smaller and faster, he strays more than the female. In the spring the male sets out to breed. That is when mating behavior can be observed, the male firmly clutching the female's back. After mating, the female builds a cocoon of white silk under a curved leaf and deposits her eggs. She stays and watches until the young have emerged and the helps them out by opening the cocoon. If the young are not eaten, they prosper all summer and spend the winter in their nursery. Because they are of different size and morphology, males and juveniles are not stealthy hunters; they pursue their prey on the run.

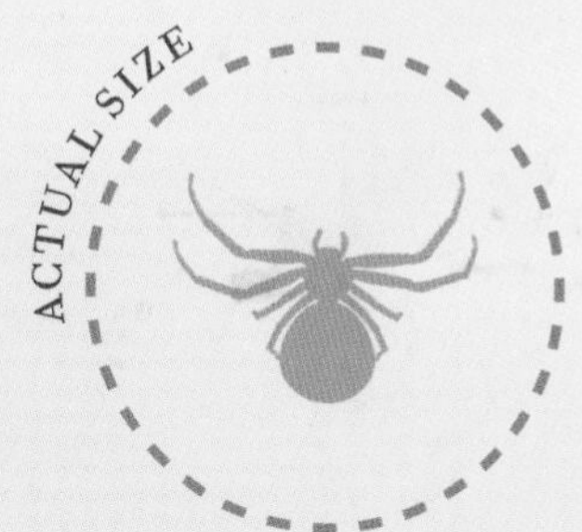

Misumena vatia
(Clerck, 1758)
Size:
Female: 9–11mm, Male: 3–4 mm

Common Domestic Fly

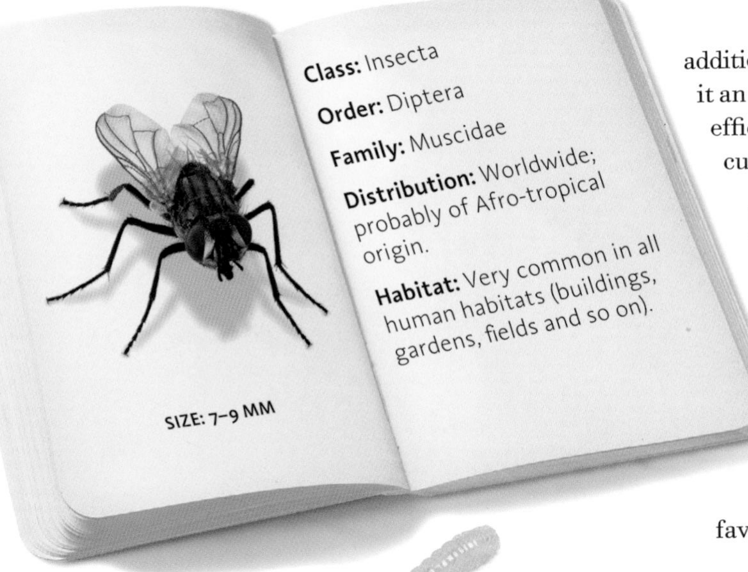

Fly larvae (maggots) and pupae

Origin of the name

This is the true fly as we all know it.
Its genus name *Musca* ("fly," in Latin) attests to this. Its species name, *domestica*, comes from the Latin *domus*, meaning "house." This extremely common species exists in constant contact with humans and lives where humans live, including their homes.

Secrets of an outstanding acrobat

Whoever has tries to hand-capture a fly in midflight knows that this requires incredible speed, concentration and dexterity. The fly will take off suddenly, spin and swerve to land a few feet away on the sleek surface of a window pane or on the middle of the ceiling. How does it do this? Although it does not appear to fly very fast (4.5 mph/ 7.5 km/h compared to 30 mph/ 50 km/h of some damselflies), the Domestic Fly's first pair of two wings beat at the extraordinary frequency of 200 times per second. The second pair of wings is transformed into two tiny appendices, "stabilizers" that detect the smallest variations in direction. The house fly's unpredictable and erratic flight pattern is an excellent defense mechanism against predators. In addition, its large eyes, with their 400 individual facets, give it an excellent 360-degree field of view, which is much more efficient than ours. Lastly, each leg is equipped with two cushions covered with hundred of silklike sticky hairs.

Lightning-fast development

The Common Domestic Fly makes its appearance in March on our windows and in our attics. After mating, the female lays 100 to 150 eggs on moist organic matter (detritus, rotten fruit, cadavers and excrement). The eggs will hatch after one to three days of incubation (though it may only take seven to eight hours on a hot summer day!). If food is abundant and weather conditions favorable, the tiny maggots complete their development (consisting of three molts) in three days! The maggot then ceases to feed, stops moving and transforms itself into a pupa, a kind of tiny brown sarcophagus encasing the developing nymph. The adult emerges three to four days later through a previously cut opening that gives way to the pushing pressure of the adult's enormous head, which is swollen with blood (photo below). Once it has emerged, the head resumes its normal proportions. The fly is then ready for takeoff.

During the summer, the female lives for only three weeks, though she lays eggs many times to reach an impressive total of 1,000 eggs. In Europe, we can observe up to five generations of this insect per year. Some individuals of the last annual generation hibernate in sheltered areas (attics, caves and stables).

The emergence of adults

GERMS ON THE MOVE

The very mobile Common Domestic Fly can go from a trash cans to cadavers of animals, from industrial effluents to dung piles ...then finds itself in our kitchen. As soon it discovers a convenient food source, it lands and positions its "trumpet." Its saliva is released and partially digests the surface of the food matter. The fly then aspirates the liquid. It can easily spread foul pathogens in this way. This fact alone justifies banning this insect from food preparation areas.

Cleaning its wings

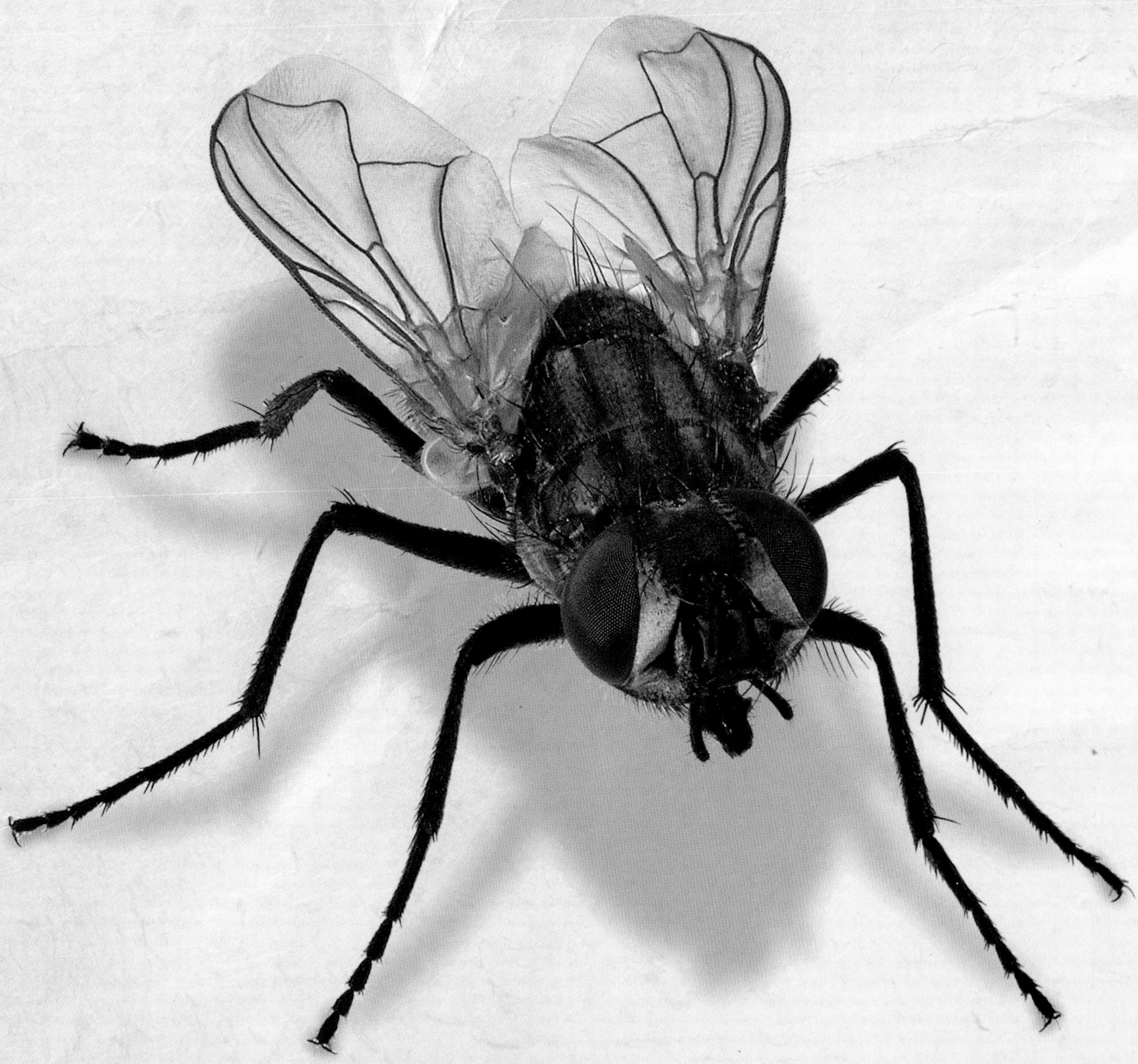

Musca domestica

Linnaeus, 1758

Size: 7–9 mm

Undertaker Beetle

Nymph of Undertaker Beetle

Origin of the name

As you may have guessed, the Undertaker Beetle is Mother Nature's burial expert. Its scientific name spells out its occupation, since *Necrophoros* means "he who transports the dead to their burial sites," and *vespillo* was the name the ancient Romans assigned to the undertaker who, come nighttime, would bury the poor.

Undertaker in a clown's outfit

The Undertaker Beetle captures odors with its antennae, as do many beetles. Its powerful sense of smell allows it to detect decomposing matter at quite a distance. It usual "costume" is one of a black body transected by two orange bands on its elytra. This beetle's foul-smelling odor tells us quite a bit about its dietary habits a fact that does not encourage casual handling. This is unfortunate because we are thereby deprived of hearing the small scream – a staccato squeaking – that it emits if picked up. Tiny dust mites live on its back and take advantage of this free ride to feed on weevils. When taking off in flight, the Undertaker Beetle has a very unique way of lifting its elytra: its beats them face to face in the center of the deployed wings. Although several species resemble it, *Nicrophorus vespillo* can be identified by its antennae, which terminate in red "pompons," and its highly curved posterior legs.

First class burial

Males often get together to exhume a cadaver whether it be a bird, a shrew or a mole. Females, attracted by the males' pheromones, soon make their appear-ance at this macabre scene. Since time is limited, the "undertakers" get busy around the remains. These tiny insects, endowed with exceptional strength to carry their large prey, go about their task and quickly dispatch the cadaver. They dig into the ground and excavate soil from each side of the "grave." As they proceed, they cut through any plant roots that get in their way and little by little, the prey's lifeless body disappears underground.

This whole task takes a few hours as the laborious digging chore continues underground. During this burying ritual, the female chooses one partner and chases away other suitors. The lifeless prey, now covered in dirt, is slowly stripped of its fur or feathers by the insects. The male then leaves the funereal chamber and the female proceeds to lay about ten yellow, elongated eggs. The eggs soon hatch and the nurturing that the mother provides to her offspring is astonishing: she even feeds them little pre-chewed "meat" balls that she has prepared with great care.

On a mole cadaver

As obstinate as an Undertaker Beetle

Several known insects, such as staphylins, dung beetles, flies and some spiders, also participate in the elimination of cadavers. The Undertaker Beetle, however, is undoubtedly the one doing the most remarkable job of cleaning up. Several naturalists, fascinated by this insect's determined behavior, have made a game of putting obstacles in its way, including spearing a mole with a stick, attaching the cadaver to a string, or hiding it under a rock. These little beetles were not to be deterred. They pushed the lifeless body with their legs and pulled it toward a better spot, amputating part of it if needed. Nothing could stop the Undertaker Beetle!

An adult and larvae

ACTUAL SIZE

Nicrophorus vespillo

(Linnaeus, 1758)

Size: 15–18 mm

Thick-legged Noble Flower Beetle

Female of this specie

Origin of the name

The genus' name *Oedemera* literally means "swollen thigh" (in Greek), a reference to the enormous femora of the males; *nobilis* means "noble." The adjective refers to the superb metallic markings of this species.

Females love big thighs

The Thick-legged Noble Flower Beetles are among the most abundant and widespread of coleopterans during summertime. It can be seen from April to July on sunny, warm summer days on a variety of flowers (ombelliferous plants, grasses, plantains, bindweed, poppies, centaury and cruciferous plants), from which they feed on the pollen. Because of this, this insect can be considered a valuable pollinator.

This beetle's very prominent eyes and long, thin antennae remind us of those of the Great Capricorn Beetle (p. 70). Their elytra, lacking chitin, are quite soft; they are narrow, tapered and spread out at their extremities so as to cover only a portion of the abdomen and wings. The males differ from the females with the swollen anterior portion of their legs. The femur reflects colors of copper or gold in the sunlight. This is a surprising characteristic, given that this species cannot jump (it flies quite well though). The females have much narrower femora. We have yet to understand this sexual dimorphism.

Discreet larva

The Thick-legged Noble Flower Beetles mate on flowers, and then the female crawls to the ground to lay her eggs. The larva digs tunnels in the decomposing wood or in the dead branches of certain plants, notably Compositae, Spanish broom plant and thistles. This larva displays typical coleopteran traits: a white, elongated, cylindrical body with legs, and a chitinous brown head. It transforms into an adult in a small, previously built loge, and only leaves this shelter the following spring.

Omnipresent though still misunderstood

More than 50 species of this beetle have been counted in Europe and North Africa. The red-brown or yellow-brown *Oedemera podagraria* is another very common species that appears on flowers a bit later in the season than the Thick-legged Noble Flower Beetle. This insect is encountered in coastal regions up to altitudes of 8,000 feet (2,500 m); from warm Mediterranean zones to the colder peat bogs. The *Oedemera croceicollis*, with a blue body and a yellow-orange thorax and forelegs, prefers to live in swampy areas. Generally, the multicolored species are found in southern regions.

BITTER DEFENSE MECHANISM

Following the example of the Cockchafers (p. 174), the Thick-legged Noble Flower Beetles defend themselves against predators by producing bitter and toxic substances, analogous to the cantharidine produced by the Spanish Fly (*Lytta vesicatoria*). Since these substances have an effect opposite to that of pheromones (which attract sexual partners), biochemists have classified these as "allomones." The Thick-legged Noble Flower Beetles – like cockchafers do not appear to produce enough allomones to cause the fearsome burns produced by the Spanish Fly.

Male on an orchidae

N.°
Oedemera nobilis
(Scopoli, 1763)
Size: 8–10 mm

Blue-winged Grasshopper

Blue-winged Grasshopper larva

Origin of the name

This species is commonly known as the Turquoise Grasshopper, the Blue Grasshopper, the Bluish Grasshopper or the Blue-winged Grasshopper. The genus name *Oedipoda* comes from the Greek and means "swollen foot." The remarkable turquoise-blue color of its wings completely justifies the species name *caerulescens*, "bluish" in Latin.

A red twin

The Blue-winged Grasshopper is hardly noticeable on the ground, due to the marbled elytra that cover its hindwings. When it takes off, however, it reveals its glorious wings with their bright blue highlighted in black. This trait can either deter or attract predators; the color gets attention from the predator for a few seconds during flight and then the insect seems to disappear as it lands. The Red-winged Grasshopper, *Oedipoda germanica* (opposite), is similar in all matters to the Blue-winged Grasshopper, with the exception of the color of its wings a brilliant red.

Red-win Grassho

Adapted to hot and dry areas

As with all grasshoppers, the Blue-winged Grasshoppers are vegetarians and feed mostly on grasses. They also appear to like dryness and heat. They are frequently encountered in rockeries, lawns and dry moors, dunes, quarries or industrial wastelands. Adults can be seen mostly in August and September.

In the mating season, the Male Blue-winged Grasshopper, visibly smaller than the female, conducts a short courtship by raising and lowering his posterior femora to seduce the female. Unlike the males of most of grasshopper species, in particular, the Meadow Grasshopper (p. 76), this one is a mediocre singer. Once in a while it emits a brief buzzing sound, almost inaudible to humans, by rubbing the lower surface of its rough "thighs" against a prominent venation on the edge of the elytra. The fertilized female digs a small hole in the ground with her abdomen, which stretches to an astonishing length. She then lays about 10 eggs, wrapped in a protective layer of foam that dries within a few hours. Once done with egg-laying, she gently taps the opening to obscure its presence and render it invisible to those in search of fresh eggs! The eggs remain in the ground for the entire winter, and the larvae (above, left) hatch the following year. Like all other orthopterans, the Blue-winged Grasshoppers are heterometabolic, meaning that the larvae, after hatching, resemble adults (though a miniature, wingless version) and then they progressively develop through successive molts.

INVISIBLE!

The Blue-winged Grasshopper has "cryptic" (from the Greek, meaning "hidden") body markings that render it nearly invisible in its environment. The marbling and dark banding of the elytra distort the grasshopper's contours. The elytra appear on a colored background, which varies according to the color of the ground color on which the insect developed: brown, grey, pinkish, yellowish, nearly white or black (photo). The whole constitutes a perfect "tone on tone" camouflage in respect to the surroundings (specialists refer to this as "homochromy").

Oedipoda caerulescens

(Linnaeus, 1758)

Size: Female 22–28 mm

Male 15–21 mm

Oniscus asellus

Common Woodlouse

Frolicking on moss

Origin of the name

Oniskos means "ass" or "donkey" in Greek, and *asellus*, means "small ass" in Latin. One wonders why Linnaeus would chose a double naming (a seemingly "ass-inine" choice!). Could it have been the grey color, or perhaps the antennae shaped like big ears? The common names assigned to this group of animals often allude to pigs: thus the name *sow beetles*. Before being shocked by this analogy, let us remember that pigs in days gone by were not pink and smooth-skinned, but rather hairy and grey.

The woodlouse — better than hygrometer

We can rarely observe this nocturnal beetle. However, even the smallest of gardens harbors millions of Common Woodlice, often hiding in walls and under large boulders. Their crustacean-like skin is not watertight, which can lead to rapid dehydration in sunlight. It makes sense that they prefer the coolness of nighttime and that they have the uncanny ability to precisely detect the amount of humidity in their environment. The Common Woodlouse is recognized by the little white or yellow spots that lend brilliance to its otherwise grey appearance.

Tomorrow I remove the front!

You can find the Common Woodlouse year-round under rocks or boards laid on the ground, or in compost piles. If winter temperatures dip too low, it will remain inactive. Should we encounter one of these Woodlice with a distinct shade difference between the front and back body portions, it is because it has just molted. Molting occurs in a two-step manner, beginning at the posterior end of the animal, which then appears much lighter than the anterior end). The Woodlouse's main predators are also nocturnal, and include toads, slow worms, and the rare spiders that can bite through its hard outer skin.

Stomach aches? Take three woodlice and off the bed!

The Woodlouse's skin, very rich in calcium carbonate, has been used since ancient times to soothe stomach aches. It is not surprising therefore, that powdered beetle was once a permanent fixture in well-stocked medicine chests.

Molting Woodlouse

TERRESTRIAL CRUSTACEANS

Let's count the legs: Woodlice do not have six like the insects, but 14. These are the only crustaceans fully adapted to a terrestrial existence: their fertilization is internal and the females lay their eggs in a moist ventral pouch, the marsupium. Once their larval development is finished, the tiny baby Woodlice, looking similar to adults, leave the pouch. These animals have made only one concession to their original aquatic milieu: they simply cannot survive in very dry environments. They should not be confused with the Millipedes (p. 128).

ACTUAL SIZE

N.°

Oniscus asellus

Linnaeus, 1758

Size: 15–28 mm

ORYCTES NASICORNIS

Rhinoceros Beetle

The Rhinoceros Beetle larva

ORIGIN OF THE NAME

The Rhinoceros Beetle obviously owes its common name to the large cephalic horn of the male. The word *Oryctes* comes from the Greek, meaning "burier," while *nasicornis* literally means "who has a horn on its nose."

A RHINOCEROS IN EUROPE

Built and leathered like an armored vehicle, the Rhinoceros (along with the Stag Beetle, p. 162, and the Ground Beetle) is one of the largest coleopterans in Europe. This beautiful insect was once found in large numbers in old oak tree groves but now is less common.

The male is immediately recognizable by its long cephalic horn, which is reduced to a small tubercle in the female. Because these homebodies are not very active, they are often not easy to spot, but they do make a little rasping sound by running their elytra against their abdomen.

ATTENTION: THE FLIGHT OF THE RHINOCEROS

Rhinoceros Beetles appear between the end of March and June and live for a few more months till fall. Males can be observed at dusk, flying noisily in search of a female companion. They often fall prey to nocturnal predators. Following in the path of the male stag beetles with their spectacular mandibles, this beetle uses the power of its horn to rebuff rivals. Females are attracted to light.

ENORMOUS LARVAE

The Rhinoceros' curved larva is an enormous white worm that can reach a length of 2 to 4 inches (6 to 10 cm). It develops in humus, compost, vegetable debris, decomposing trees, or on piles of humid sawdust from sawmills. In the past, it was also found in piles of leather scraps in tanneries. In Spain we have found Rhinoceros larvae in the trunk of a date palm tree. In Southern areas, its large size does not protect it from an equally large parasitic Hymenopteran: the Mammoth Wasp (Scolia flavifrons). This huge wasp digs deep into the ground to find the host larva and then immobilizes it with a sharp blow and lays an egg inside its body. After hatching, the small wasp larva pierces the skin of the huge Rhinoceros larva, and devours it from the inside out. After three to four years of larval existence (more if food supplies are scarce), the Rhinoceros larva surrounds itself in the fall with a shell of soil and vegetable debris, in which it will first transform itself into a nymph, and then an adult. The latter does not emerge from its loge until the following spring or summer.

A HERCULIAN STRENGTH

Research conducted on the North American Rhinoceros Beetle (*Xyloryctes thestalus*) has shown the colossal strength of these insects. Even when laden with a weight 100 times greater than its body mass, the Rhinoceros still manages to move! Furthermore, it displays a surprising capacity to conserve energy: for example, when it carries a weight 10 times its own weight, its energy output is only doubled! In comparison, a man transporting a load 20 percent larger than his body mass must expend 20 percent more energy.

The female (below) does not have a horn

ACTUAL SIZE

Oryctes nasicornis

(Linnaeus, 1758)

Size:

20–40 mm

Hermit Beetle

Origin of the name

Also known as the Russian Leather Beetle, The Hermit Beetle owes its genus name *Osmoderma* (from the Greek, *osmê*, "odor" and *derma*, "skin") to the surprising odor of prunes it emits! As for its species name, *ermita*, this reflects this beetle's solitary and secretive mode of existence.

An insect that stopped a highway

Although this insect is just over an inch long (3 cm) and not very noticeable, just its presence was enough to stop the bulldozers building a major highway in the Sarthe region of Southern France. The highway was detoured in order to avoid destroying a large number of trees harboring the Hermit Beetle. This species now commands the highest protection level, both in Europe and in France.

Discreet host in old tree trunks

With its black-brown coloration and brilliant purple reflections, this beetle is closely related to the Golden Cetonia (Rose Chafer, p. 74).

The larva feeds in hollowed-out tree cavities full of compost. These are created by the degradation of wood caused by fungi and xylophagous (wood-eating) insects. The female lays her eggs in the compost.

The larva of the Hermit Beetle

Following incubation, the larva begins its long sedentary life, which can last up to three years. At the end of its last winter, the larva readies itself to become a nymph and builds a 1 to 2 inch (2.5 to 4 cm) compost shell. An adult will emerge the following spring, shortly after metamorphosis. If the exterior temperature is favorable, the adult will surface from their composted place of birth. Though they can fly, they do not go very far. The rare individuals that choose to leave their birth cavity do not venture farther than about 30 feet (9 m), or at the very most, 330 feet (100 m) away.

A rough habitat

Populations of Hermit Beetles require a sufficient density of hollowed-out trees for their survival. Depending on the variety, these trees can be from 50 to 150 years old. Unfortunately, frequent harvesting of trees leaves very little time for them to grow old. Strangely, the Hermit Beetle still exists in farmlands, thanks to a rural tradition of pruning trees to obtain firewood or food for cattle. This has led to older trees, in which cavities can develop. In many places, a species that originated primarily in forests is now living principally in farmlands. Unfortunately, because of land consolidation and intensive cultivation, more than two-thirds of these lands have disappeared in the past 50 years. The unusual biotope that this insect chose to live in is no longer widely available.

An "umbrella" species

While enacting laws for the protection of the Hermit Beetle, France is also including a wide variety of narrowly linked species under this protective umbrella: numerous insects, small rodents, small birds of prey, bats – all species that they are trying to conserve. It is about time that humans and all other species fall under the same umbrella! The farmlands were there to protect us against wind, drought and erosion. They also shelter a multitude of beneficial insects and provide shade for farm animals.

Osmoderma eremita

(Scopoli, 1763)

Size: 24–35 mm

ACTUAL SIZE

Green Shield Bug

SIZE: 10–14 MM

Class: Insecta

Order: Hemiptera

Family: Pentatomidae

Distribution: Europe (more widespread in central and southern); introduced to the United States in the southeastern states.

Habitat: In gardens and forests on diverse plants and trees.

Two at larval stage and one molting

ORIGIN OF THE NAME

The origin of the genus name is obscure. The species name is clearer: *prasina* means "green" (the green color of a pear) in Latin.

SEASONAL COLORATION

The Green Shield Bug only merits its name for a certain time of the year: it is green in summertime and goes brown just before hibernating, but returns to a brilliant green in springtime. This change in color enables this insect to move about unnoticed. We encounter it from April to October in a variety of biotopes – wherever there are trees and shrubs – including meadows, gardens, hedges and the edges of forests. It often lives on ombelliferous and herbaceous plants. Both adults and juveniles not only suck up the sap but also attack the larvae of other insects. Like the adults, the juveniles are green. In June, females lay about 200 eggs in groups of 20 to 30, on the underside of leaves. The females often stay close to their eggs for a few days and then abandon them. At the end of a two- to three-week incubation period, the transparent, developing embryos reveal the small red eyes of future larvae. Once the larvae have hatched, it takes five to six weeks for them to become adults.

FEARED IN ORCHARDS

Because they are numerous and can adapt to several types of plants, the Green Shield Bugs can cause much damage to orchards, especially those in proximity to vast forests or to herbaceous plants (either wild or cultivated varieties). This insect has a preference for young pear or apple trees. They pierce the fruit for nourishment and inject it with saliva, which causes degradation of the fruit's skin and hardening of the flesh.

ALMOND PERFUME

The Green Shield Bug often chooses to live in houses in the off-season. It can put out a very strong odor, which is passed on to the fruit it pierces (raspberries for example). This odor is considered quite disagreeable by many, though to some it is a rather pleasant smell of bitter almond. It's all a matter of personal olfactory sensitivity!

The less common species, *Palomena viridissima*, is identical to the Green Shield Bug species (*P. prasina*) with the exception of a barely perceptible detail: the third segment of its antennae is clearly smaller than the second one, whereas they are nearly all the same length in *P. prasina*.

THE GREEN SHIELD BUG, WHAT A SMELL!

The Green Shield Bug has specialized glands on the side of its thorax, close to the base of its posterior legs, which produce a complex "cocktail" of volatile smelly substances. These include a mix of aldehydes, which are responsible for the characteristic odor of these insects as well as aromatic hydrocarbons and turpentines. These are formidably efficient, even at very low concentrations. Not all bugs have this unpleasant odor – some smell of apples, hyacinth or licorice!

Newly hatched larvae

The long stinging rostrum

ACTUAL SIZE

Palomena prasina

(Linnaeus, 1761)

Size: 10–14 mm

Panorpa communis

Common Scorpionfly

The female does not have an appendage in the form of a scorpion tail

Origin of the name

The word *Panorpa* comes from the Greek, "pan," "all," and *horpêks*, "stinger." Linnaeus refers to the fact that the Scorpionflies appear to be "armed" at both extremities. The tiny head, perpendicular to the rest of its body, terminates in an imposing rostrum with significant mouth parts. At the other extremity, the long raised "tail" ends in a swollen manner reminiscent of the scorpion's stinger. This odd appendage has given this insect its common name of "Scorpionfly." In reality, this swollen part consists of claspers, which are only found in the male and are used in mating.

Scorpions or vultures?

Despite their strange and worrisome appearance, Scorpionflies are perfectly harmless. They are usually found in damp, shady locations such as hedges, edges of forests and on roadsides. They are sometimes confused with the Vulgar Scorpionfly (*Panorpa vulgaris*), which has more distinct black markings and prefers sunny hedges and bushes.

We observe this insect from spring to fall, perched on vegetation, waiting for its prey. If its silhouette reminds us of a scorpion, its feeding behavior resembles more that of a vulture. This opportunistic carrion-eater feeds mostly on dying or dead insects. Scorpionflies are often seen surrounding and fighting aggressively over the carcass of a dead insect. On occasion, they may also feed on detritus of plants, fruit, nectar, or the honeydew of aphids.

Seducing males

In order to attract females, males put out pheromones from glands situated at the level of the swollen section of their abdomen. He seals the deal by offering her a special treat, an unexpected "gift" – either an insect's head or a strange, hardened column of saliva that becomes gelatinous when exposed to air. While the distracted female is busy eating her tasty treat, the male grasps the extremity of her abdomen and inserts his copulatory organ, which is shaped like the tail of a scorpion.

The female lays her eggs in the ground. The larva, resembling a hairy caterpillar, feeds on the cadavers of soft insects or larva. Following about three molts, which last about a month, the larva enters diapause for a period of approximately five weeks, after which it will be transformed into a nymph. In some species the nymph hibernates in the ground and only emerges as an adult the following spring. In other species, the adult appears at the end of summer. The Common Scorpionfly has only one generation per year in northern regions, but it can have two generations farther

Odd profile of the Scorpionfly

FLEAS AND SCORPIONFLIES ARE COUSINS

The Panorpidae Family has 350 species, most of which are found in the Far East and North America. Recent molecular biology investigations have confirmed that Fleas (classified until recently in the order Siphonapterae) belong to the order Mercopetera and, because of this, are distantly related to Scorpionflies. This is another example of how in entomology, as in other disciplines, we must be wary of judging only on the basis of appearances.

Panorpa communis
Linnaeus 1758
Size: 15–20 mm
ACTUAL SIZE

Old World Swallowtail

Passage from caterpillar to chrysalis

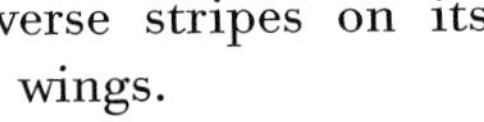

ORIGIN OF THE NAME

The word *papilio* is simply the general Latin name for all butterflies. When Linnaeus was establishing the nomenclature of the species, he was in the habit of assigning the names of Greek mythological figure to certain butterflies; Machaeon was one of the three sons of Asclépios (god of medicine). The legend is that Machaeon was also a doctor and took a seat in the Trojan Horse. The ancients attributed several other names to this butterfly, including "Big Tail Carrier" and "Swallowtail." We need to be careful when using these terms since they also refer to many other butterfly species. I personally like "Big Carrot-Eater," a name given to this butterfly for its predilection for eating carrot tops.

SUMMIT MEETING

The Old World Swallowtail is one of the largest and most beautiful butterflies. Though it was never very widespread, it has become scarcer, fallen victim to the misuse of pesticides. It can be mistaken for a close relative, the Scarce Swallowtail (*Iphiclides podalirius*), an even rarer species that is noted for the transverse stripes on its wings.

The Old World Swallowtail flies from April to September. Unfortunately, it never seems to hover in the same place, thereby depriving us of viewing its magnificent aerial display. Because of its large wingspan, this butterfly is an excellent glider. In order to meet his "lady," the male swallowtail practices a flight pattern known as "hill topping," whereby he twirls about the hilltop, fighting off other males for the best spot on the summit. The winner's prize will be to have the choice location to connect with the next incoming female.

A FAN OF CARROTS

Once she has been fertilized, the female must find an appropriate host plant to feed her progeny. The eggs are laid one by one under the leaves of ombelliferous plant species (either wild or cultivated) such as carrots, dill, fennel or caraway. At first the young caterpillars are black with a white spot down the middle. They assume their vibrant coloration after the third larval stage. I urge you to let these beautiful caterpillars munch on your veggies for a while! Your garden will survive, plus you may just have the chance to witness the metamorphosis into a chrysalis. This process goes as follows: first the caterpillar weaves a small silk belt that serves to hold it onto the host plant while it feeds, then it folds itself in half before undergoing its own transformation. Undoubtedly for reasons of careful camouflaging, the chrysalis first appears green in spring and then slowly changes to a brown color at the end of summer. There are usually two generations per year, depending on the region's climate conditions. The end-of-season chrysalises hibernate and hatch the following spring.

DO NOT TOUCH ME

When threatened, the caterpillar shoots out a kind of small forked, telescoping, red-orange horn (the osmeterium) from the top of its head (opposite). This is a small, odiferous organ that emits a very bad smell intended to ward off any aggressor.

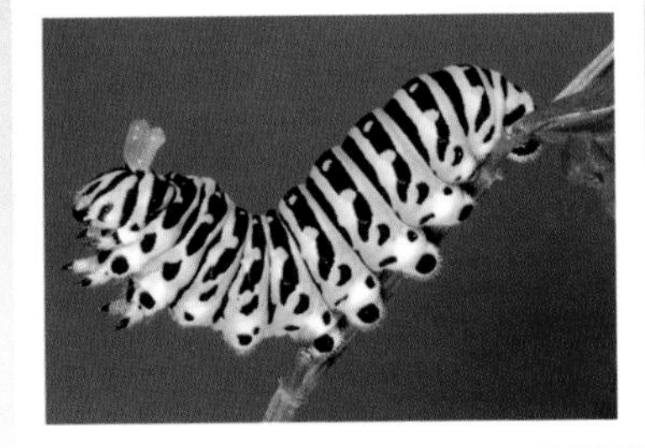

ACTUAL SIZE
N.° Papilio machaon
Linnaeus, 1758
Wingspan:
70–90 mm

Harvestman Spider

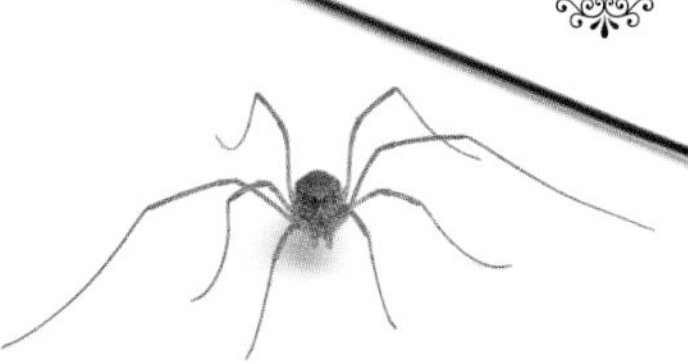

A youngster not much bigger the head of a pin

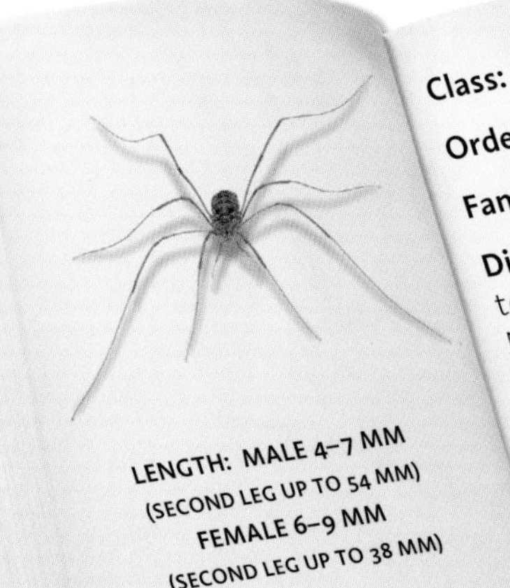

Class: Insecta

Order: Opilionidae

Family: Arachnidae

Distribution: Europe, temperate zones of Asia and North America.

Habitat: In the low-growing vegetation of woodlands, hedges, cultivated fields and gardens.

ORIGIN OF THE NAME

Linnaeus adopted a name that the Greeks (*phalangion*) and then the Romans used to designate the tarantula, a large spider considered venomous. This was a very lofty name for such a modest and inoffensive spider as the Harvestman Spider, also known as Daddy Long Legs. The species name, *opilio*, means "shepherd" in Latin, a reference to the shepherd who runs after his flock just as a Daddy Long Legs scampers about with his long appendages.

NAME THE SEVEN DIFFERENCES GAME

At first glance, with its four pairs of elongated legs and its stocky body, the Daddy Long Legs immediately conveys the image of a spider. Upon closer examination, however, we soon discover that it is very different, to the point of belonging to a distinct subclass with the following characteristics. 1. Its body is all in one piece (head, thorax and abdomen are fused), but in true spiders, the body is in two segments: the cephalothorax and the abdomen are linked to one another via a thin pedicle. 2. It has one pair of eyes (true spiders have three or four pairs), perched on a protuberance in a manner resembling a periscope. 3. Its chelicerae terminate in pincers. 4. Its two masticating legs (pedipals), with reduced claws, are like supplementary legs. 5. The second set of legs, much longer than the others, appears to have significant sensory functions. 6. It does not have any venomous glands nor does it produce silk or weave a web. 7. The male is equipped with a true penis, while true spiders use their "boxing gloves" in mating (see example on p. 106).

USEFUL ANIMALS

A nocturnal specimen like most Opilionidae, Daddy Long Legs can be seen from May to November. We can identify it by the two clear denticles at the edge of the front of the cephalothorax. The male differs from the female by the presence of a long "horn" on the chelicerae. The Daddy Long Legs spends its days searching on walls and in open tree trunks for dead insects and plant detritus. It also tracks aphids and mealybugs and other small limacea found in gardens. If threatened, it will run quickly with a somewhat gangly gait on its long, spindly legs not unlike those of a wading bird.

BEWARE OF FEMALES!

It is not unusual for the female to devour the male after mating. Contrary to a popular myth, this mating behavior is not a common occurrence, even with spiders. If this occurs, it will most likely happen at the end of summer, when individuals approach maturity. The female lays her eggs with her ovipositor in damp crevices of rockeries or tree trunks. The eggs hibernate, along with a certain number of juveniles.

The eggs of a Daddy Long Legs

THE "ARMS" FALL OFF

When we try to capture the Daddy Long Legs, we are often left with many of its legs in our hands. This process, by which an animal can sever its own limbs, is known as autotomy and enables it to escape its predators. The severed limb retains some jerky motion, which can lure and trick a predator. A severed leg cannot regenerate (contrary to what happens with the true spiders). Therefore, individuals with less than their original eight legs are a fairly common sight.

All legs

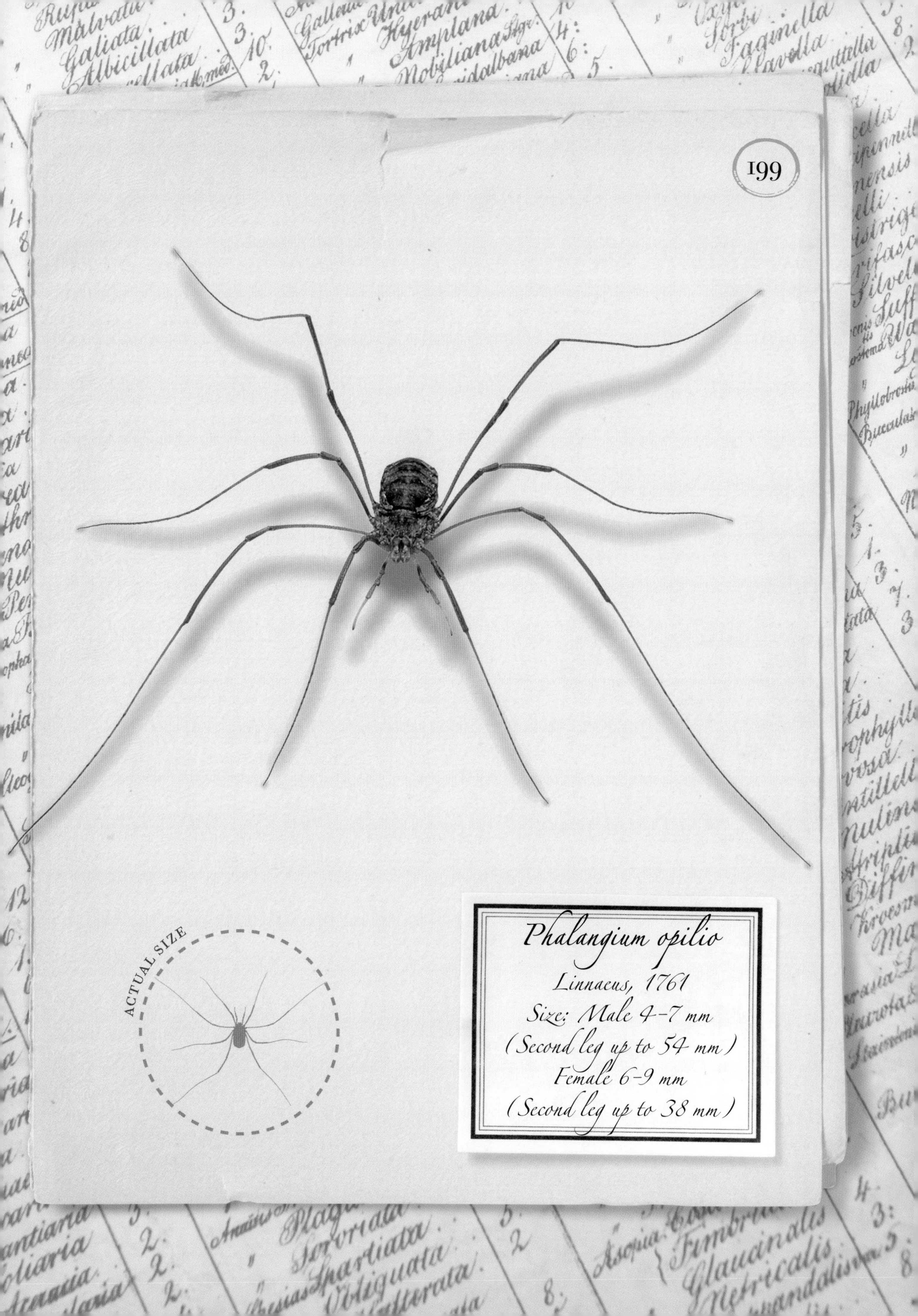

Phalangium opilio

Linnaeus, 1761

Size: Male 4–7 mm

(Second leg up to 54 mm)

Female 6–9 mm

(Second leg up to 38 mm)

Cabbage White Butterfly

Cabbage White Caterpillar

WINGSPAN: 50–65 MM

Class: Insecta

Order: Lepidoptera

Family: Pieridae, also Nymphalidae

Distribution: Originated in Europe, Western Asia and North Africa; today, it has spread to nearly all of temperate Northern Hemisphere, Australia and New Zealand.

Habitat: All open spaces where cruciferous plants grow. This is a migratory species, often seen in urban areas.

Female Cabbage White at rest

ORIGIN OF THE NAME

The White Cabbage Butterfly, one of our most common butterflies, was given the genus name of a muse, *Pieria* (born on Mount Piéros in Thessaly). The cabbage plant, the favorite snack of its caterpillar, is named *bassicus* in Latin, from which came the species name of this butterfly, *brassicae*.

CABBAGES BEWARE

The female Cabbage Butterfly has two small spots underneath her forewings; these are absent in the male (see below). They have at least two generations per year and often more, which overlap in such a way that we can simultaneously observe all the developmental stages. The female lays up to 300 eggs under a leaf, in packets of about 10. The pretty little yellow eggs are finely carved "bombs." After about a week or so, tiny little caterpillars will emerge and begin by eating the egg's chorion or shell, then they proceed to attack cabbage leaves. At first they are quite gregarious, but they become more solitary as time passes by. Their constant appetites guarantees that they will reach a length of just over one inch (3 cm) in the space of two to three weeks.

The male: no round spots on the forewing

DEFENSELESS CABBAGE

Cabbages have physical weapons – a waxy layer covering the leaves and chemical weapons such as glucosinolates – which are responsible for the characteristic cabbage odor that repels most insects (and some people!). Unfortunately, these weapons are inefficient in repelling the Cabbage White Butterfly or its caterpillar. The butterfly, far from being deterred by the cabbage's chemical arsenal, uses it to its own advantage while laying its eggs. In addition, with the help of a digestive protein, the larva is capable of neutralizing the toxic compounds produced by the cabbage.

APANTELES TO THE RESCUE

Fortunately, the cabbage has several precious allies: tiny parasitic wasps, notably the Apanteles (p. 96), which attack Cabbage White caterpillars. After feasting on the leaves, the non-parasitized caterpillars leave the cabbage food and climb along a wall or into a tree. They then weave a thin silk belt around their waist and are transformed into chrysalises. At this stage, another parasitic hymenopteran comes in to play – the miniscule *Pteromalus puparum*, which parasitizes only the chrysalises (photo at left). The surviving chrysalises hibernate in this state and then metamorphose into adults during the beautiful days following spring.

THE CATERPILLAR THAT STOPPED A TRAIN

The caterpillar mostly attacks cruciferous vegetables (cabbages and broccoli) but also watercress and nasturtium. The proliferation of vegetable gardens has greatly benefitted this caterpillar. Following the example of the Colorado Potato Beetle (p. 154), the Cabbage White was once a major plague. There are stories told that in Czechoslovakia in 1854, a train was actually stopped by – caterpillars! The latter had been attracted by the smell from a field of cabbages and crossed a train track in such large numbers that the train wheels, now covered in millions of caterpillar remains, began to skid uncontrollably! However, with the exception of a few farm regions, this butterfly has become much less common.

Pteromalus puparum *on a chrysalis*

ACTUAL SIZE

Pieris brassicae

(Linnaeus, 1758)

Wingspan: 50–65 mm

Flat-backed Millipede

Origin of the name

This millipede is recognizable by its pretty calcareous rings or plates, which give it a very rugged appearance. The upper portion of these plates is flat and in the shape of a rectangular hull, which is angled and delicately ornate. *Polydesmus angustus*, which literally means "with numerous fitted rings," clearly defines it. This is similar to its older name, *P. complanatus*, which means "with many flattened belts."

The millipede "wave"

This polydesmid is a multifooted member of the Polydesmidae family. In Greek, *myria* and *podos* means "very many feet." Myriapod is consequently a more appropriate name for it than the more familiar millipede, particularly since females only have 30 pairs of legs and males 31. As the class name indicates, diplopods have two (*diplos* =double) pairs of legs per segment. Unlike chilopods, like the Brown Centipede (p. 160) or the House Centipede (p. 222), the Flat-backed Millipede's legs work in a synchronized manner. Each pair of legs is raised one after another, much like the wave sport fans make in a stadium, forming a continuous flow that is fascinating to watch.

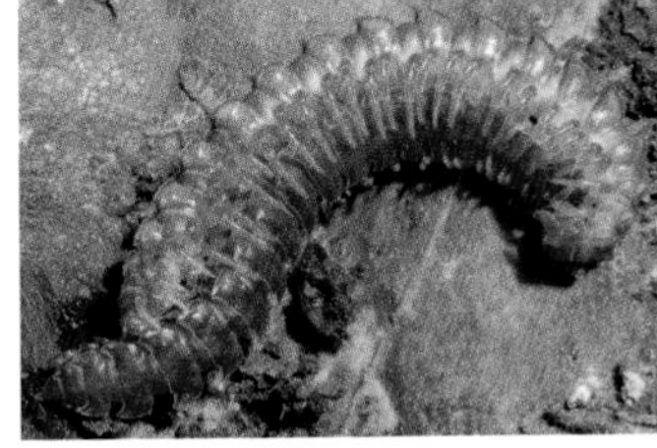
Coupling

No need for eyes

We often find millipedes in our gardens and in compost piles, where they will quickly roll up on themselves if we annoy them. However, they are most abundant in forest humus. Like all Diplopods, they primarily consume decomposing vegetation, although millipedes will also eat fruit like wild strawberries and roots. Since they live in the dark, they lack eyes but have ultra-sensitive antennae (pictured below), that are covered with multiple sensors to constantly monitor the surroundings.

In the arms of a millipede

During mating, millipedes wrap one another in their many feet. They move slowly forward against each other until the male's genital organs are located beneath those of the female. Sperm then moves along a special duct (the seminal duct), before being transferred into the female's genital opening. After laying her eggs, the female protects them in a small earthen dome, in which she creates a fine opening, through which the offspring can emerge. As with the Pill Millipede (p. 128), the young larvae only have three pairs of legs. With each successive molt to sexual maturity, as additional segment rings are added, they harden into the millipede's characteristic appearance. They spend the winter under rotting vegetation.

A POISON EXPERT

In order to defend itself, the millipede sprays a jet of hydrogen cyanide at its aggressor. How can it do this without endangering its own life? The millipede does not actually harbor the final product within its body, but stores an inoffensive precursor in its glands. Thanks to enzymes, it's not till the moment of expulsion that the toxic mixture forms.

The millipede has no eyes

Polydesmus angustus

Latzel, 1884

Size:

14–28 mm

(males are slightly larger than females)

PSYLLIODES CHRYSOCEPHALA

Cabbage Stem Flea Beetle

A related species

Every species of these beetles is usually linked to a particular family of plants. *Psylliodes affinis*, for example, favors the Solanacea (including potatoes), and *P. napi* feeds on turnips and watercress.

Damage done to hollyhock leaves

Origin of the name

The terms *Psylliodes* is of Greek origin, meaning "a type of flea." The species name, *chrysocephala*, signifies "with a golden head," due to its metallic sheen and dark background color. This beetle is also called the Big Winter Beetle.

Jumping like fleas

The genus *Psylliodes* encompasses about a hundred species in Europe and Asia. These miniscule insects, active in sunlight during warm periods, cannot be captured. Thanks to their highly developed thighs, they can execute huge leaps.

The farmer's dread

Cabbage Stem larvae grow on a large variety of plants. Members of the cruciferae or brassica families are especially favored. This beetle attacks cabbage plants, rape seed and cauliflower. It has very much profited from monoculture agricultural practices, since in nature it is no more harmful than other insect species.

Raiding cabbage fields

The Cabbage Stem Flea Beetle can cause serious damage. Adults can completely destroy young seedlings and seriously weaken fully grown plants. On the other hand, when larvae attack plant stems and thereby retard growth, the infestation is much reduced.

In any case, there is only one generation of Cabbage Beetle per year. Adults appear in spring and gnaw on leaves, leaving small holes in them. They likewise attack the flat fruit pods of cruciform plants. In the summer, the larvae enter diapause (dormancy) at the edge of the fields, providing a short respite for growers! After re-awakening, the beetles attack fresh seed and young seedlings. At the end of 10 to 15 days, the females deposit between 70 and 150 eggs at the base of the plants. After eclosion, the young larvae climb the branches of plants and dig into the petioles of old leaves, which causes them to fall. The larvae then move into the cores of branches and finally gnaw their way into the terminal buds. Nymph formation takes place in the soil. The adult beetles emerge at the beginning of autumn and pose a threat to the hibernating seedlings.

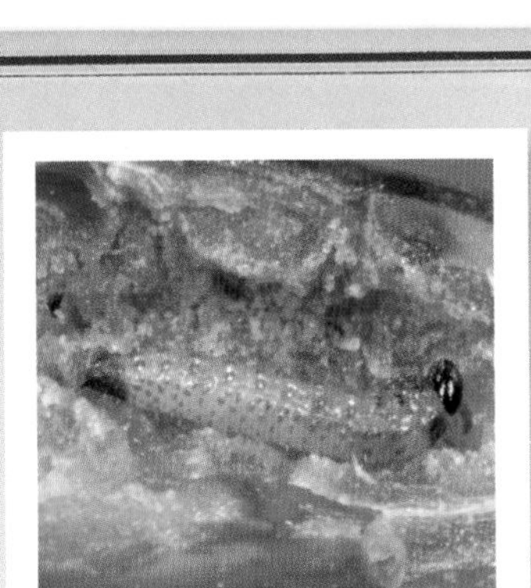

Cabbage Stem Flea Beetle larvae

HOW TO FIGHT THEM BIOLOGICALLY

Invasions of Cabbage Stem Flea beetles can be efficiently prevented without using tons of pesticides. Water the tender leaves with cold water in the morning and early afternoon (these beetles do not like cold showers!). Throw some ashes on the leaves and at the base of the plants. Plant clover and chrysanthemum as companion plants in the garden. The beetles detest them and they will turn away toward other plants they like, such as radishes and mustard.

ACTUAL SIZE
Psylliodes chrysocephala
(Linnaeus, 1758)
Size: 3–4.6 mm

Pyrochroa coccinea

Cardinal Beetle

The Cardinal Beetle larva

Origin of the name

The common name refers undoubtedly to the red coloration of the Cardinal Beetle – a color that symbolizes its namesake's lifelong devotion to the Catholic Church. This is not an original name, since it also refers to a small North American bird (genus *Cardinalis*) and to a small fish much prized by aquarium-keepers (genus *Tanichthys*). The Latin name, however, sheds a different and more precise light on this: *Pyrochroa* literally means "the color of fire" and *coccinea* means "vermillion." The Cardinal Beetle is also sometimes designated under the more scientific name of Scarlet Pyrochre.

Extra-flat larvae

Although even the adults are not very round, the larvae, on the other hand, are very flat. They are A beautiful yellow color, they are over 1 inch (3 cm) in length and live under partially detached pieces of bark, where they chase away the larvae of other insects and feed on wood fibers. These very vigorous larvae use the two spines at the end of their abdomen to lift the bark and free it from the underlying woody tissue. If we lift the bark of dead trees, we can see them quickly scurrying to hide. Due to their vague resemblance to Millipedes (p. 160), they can arouse repulsion and fear, which is perfectly justified! It would be prudent to replace the bark pieces and leave this insect at peace. After two to three years in the larval stage, the change into nymph occurs under the bark in a circular chamber (loge). The adult emerges 10 to 15 days later.

Hyperactive adults

The Cardinal Beetle is a very active coleopteran in the sun; it lives in May and June in woodlands, clearings and hedges. Females come to lay their eggs under the bark of felled trees or in old worm-eaten stumps that teem with a large quantity of insects. Its deep red color and its rather skittish behavior makes this an easily trapped insect, either in flight or on the ground, where it lives among low-growing vegetation. The male can be identified by its long comb-like antennae. The female's antennae are more dentate in shape.

Super-easy identification

Two other easily identifiable species of Pyrochroidae are also found in our regions. *Pyrochroa serraticornis* (photo below) is distinguished by its red head (instead of black), while *Schizotus pectinicornis* is red-orange in color, often with a black spot on its pronotum.

A USEFUL INSECT

The Cardinal Beetle plays a very positive role in controlling the populations of xylophagous insects (that develop in wood) of forests and hedges, such as the European Spruce Bark Beetle (p. 148). When the latter proliferate, the hungry Cardinal Beetle larvae also grow in numbers as they prey on the little invaders. When prey is scarce, the Cardinal larvae do not hesitate to eat one one another. Since this gorgeous species beautifies our forests through its unusual coloration, it deserves our attention.

Two views of a closely related species: Pyrochroa serraticornis

ACTUAL SIZE

Pyrochroa coccinea

(Linnaeus, 1761)

Size: 15–18 mm

PYRRHOCORIS APTERUS

Firebug Beetle

Young Firebugs

SIZE: 9–12 MM

Class: Insecta

Order: Heminoptera

Family: Pyrrhocoridae

Distribution: Europe (rare or absent on islands), Asia and North Africa.

Habitat: Very common in urban areas as well as farmlands, at the foot of walls or under old lime trees.

The firebug's technique

Firebugs mate in springtime. The males, which are generally smaller than the females, stick to their partners in a back-to-back manner. These couplings can last for several days. Obviously, being hooked on to one another in this back-to-back fashion greatly limits the daily activities of these insects. As a result of this unusual "sexual technique," the stronger of the two partners (often the female) must drag the other one around. Following egg laying, the young will hatch and develop between June and October. Like most other beetles, the young resemble adults and share the same lifestyle. In fact, these insects are "heterometabolites;" that is to say, they undergo incomplete metamorphosis.

Origin of the name

The bright colorations of this insect have given it many common names: soldier, guard, Swiss guard and so on. These names seem to refer to the red-and-black colors once found on guard uniforms. Its scientific name, *Pyrrhocoris*, means "bug of fire" in Greek – a reference to its brilliant color. The species name, *apterus*, on the other hand, means "wingless," although a small portion of individuals retain vestigial hindwings and others are even equipped with long wings. However, despite the length of their wings, Firebug Beetles are unable to fly.

Don't fear the firebug

Firebugs have the habit of assembling in large groups – often hundreds of individuals. While this could lead to fear of ravaging infestations, in reality, Firebugs cause very little damage thanks to their varied diet (known as polyphagous, meaning they feed on many different foods). It uses its piercing and sipping mouthparts, which act as a rostrum to feed, and is particularly fond of fallen lime trees as well as primulas (purple variety), dead insects and the eggs of other insects.

The south-seeker

The Firebug Beetle is a heminopteran, that is to say a "bug." It emits a substance with a repulsive smell as a defense mechanism. Very early in the season, as the sun warms the atmosphere, we begin to see the appearance of small red "spots" on the trunks and at the base of lime trees (lower right photo). These are actually adults from the previous year that have hibernated in the shelter of the bark or under rocks; once they emerge from their winter hiding places, they regroup under the warm sun. We can also see them on sun-baked walls and stones, both in rural and urban areas. This behavior has also given the Firebug Beetle another common name that has nothing to do with its bright coloration: the South-seeker.

A FEARSOME MASK

The Firebug Beetle is one of the most popular and well known of all insects. Obviously, though, it does not even try to blend in! The spots on its back look like a drawing of an African mask (photo to the right). Children love to look at them upside down, as they then look like the head of a man with big eyes and a large nose. However, as far as the insect is concerned, these markings are not a disguise but serve more as an early warning to predators that they taste terrible!

An African mask on their backs

Firebugs love company

ACTUAL SIZE

Pyrrhocoris apterus

(Linnaeus, 1758)

Size: 9–12 mm

Common Red Soldier Beetle

Mating on flowers

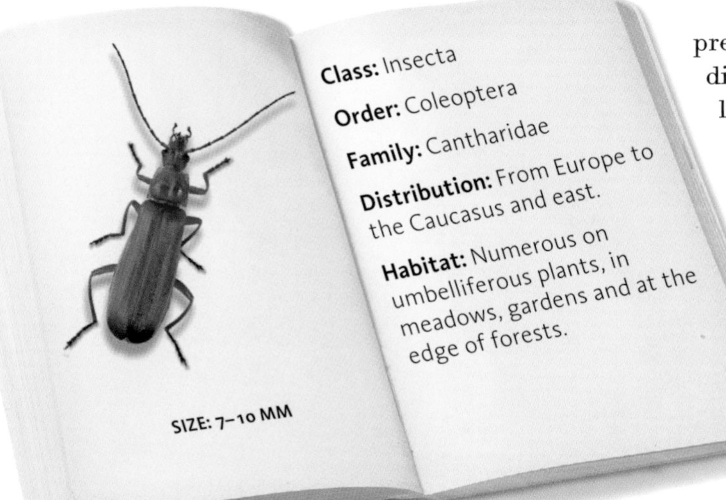

Origin of the name

The genus name *Rhagonycha* literally means "cracked fingernail," in reference to the heart-shaped design on this insect's tarsi; and the species name *fulva* indicates its "tawny" coloration. Its common French name, *téléphore*, is derived from *Telephorus* (the old genus name of this species), from *tele*, meaning "from afar" and *phorus* for "carry," in reference to the fact that these insects are often extremely widespread. This beetle is also called the Yellow Rhagonite.

Widespread on flowers

The Red Soldier Beetle is omnipresent in the summer in meadows and at the edge of trails. You find them most often on umbelliferous flowers in the company of bees and butterflies. Active only during the day, they move like flies from flower to flower looking for pollen, but don't hesitate to augment their diet with aphids when the occasion presents itself. Females are easily distinguished due to their much larger abdomens as compared to males. Mating takes place on umbelliferous flowers, often with many pairings at the same time. The female subsequently deposits her eggs on the ground. The yellow-brown larvae, which have a velvety texture, spend the winter under rocks or in underground tunnels and metamorphose into nymphs the following spring.

Malacoderme: soft-skinned animal

As recently as the 20th century, the name "malacoderme" was still used to describe a heterogeneous mix of Coleoptera families because, like the Cantharidae, they all have soft integuments. Consequently, after they die, their elytra tend to shrivel up when they become dried out. Beetle collectors have largely avoided studying these insects – mainly because they show poorly when displayed in collections – and have focused instead on scarabs and other beetles of "purer" coleopteran lineage.

Beautiful biodiversity to study

Related to Glow-worms, (see p. 152) the Cantharidae differ significantly in that they do not emit light and that females resemble males. Another very common species of soldier beetle, *Cantharis fusca*, is easily spotted since it is larger and has a red-colored thorax; and we often see couples on flowers. Many other species of Cantharidae, however, have gone largely unnoticed in our regions – especially those at higher elevations – to the point that we really don't know how many species there actually are! Will anyone take on this task?

Ready for takeoff

"Snow worms"

Red Soldier Beetle larvae live under rocks, in litter or in caves, where they feed on small mollusks. On sunny winter days, however, they occasionally leave their shelters and come to the surface. They can appear in considerable numbers on the snow at these times. Impressed by such an event, ancient people gave them the name "snow worms." Their behavior was sometimes taken as an omen or a foreteller of catastrophic events. While this cold-weather phenomenon might have been common in the past, it seems to be quite rare these days.

Red Soldier Beetle larva

ACTUAL SIZE
Rhagonycha fulva
(Scopoli, 1763)
Size: 7–10 mm

Red Palm Weevil

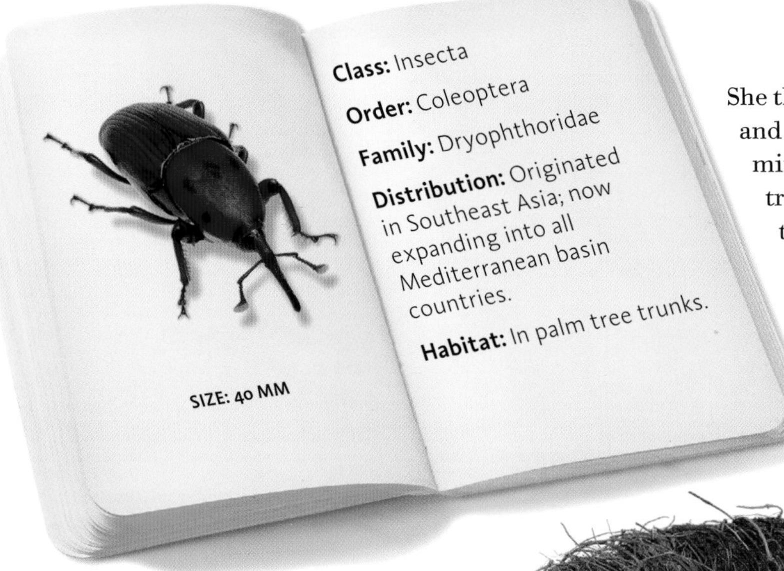

Cocoon wrapped in palm fibers

ORIGIN OF THE NAME

The scientific name of this large, ravaging insect comes from the Greek *rhyncho*, for "snout" or "bird beak," and *phor*, for "protruding" – referring to its extended rostrum. Its Spanish common name, *Picudo rojo*, is also tied to its characteristics.

A NEW INVADER

This is France's largest insect pest. While it may be discreet, its exotic appearance is hard to miss. Originally from Southeast Asia, the Palm Weevil has invaded most of the Middle East as well as Egypt, Spain, Italy, Turkey and the Canary Islands. Arriving in France in 2006, this insect is cause for much concern to agricultural authorities. It has already been found in Var, the Bouches-du-Rhône region, l'Hérault, the eastern Pyrenees and Corsica, profiting from our increasing habit of growing palm trees in our gardens and towns. In Europe it attacks mainly palm trees of the genus *Phoenix*.

A HUGE APPETITE

Irresistibly attracted by the scent of palms, the female looks for the soft base of or a wound in the trunk and digs a 5-mm-wide hole into it with her rostrum. She then turns around, deposits an egg with her ovipositor and repeats the process a bit farther along. In a few days, minuscule larvae emerge and feed on the surrounding tree pulp. As it grows, the larva begins a journey that takes it to the very core of the palm tree. This soft portion is rapidly consumed and becomes a cavity filled with shredded fibers that undergo fermentation and raise the internal temperature up to over 100°F (38°C) – a little incubator ideally suited for this tropical invader!

About 96 days later, the now 2-inch (5 cm) long white larva begins the process of nymph formation. It moves to the periphery of its cavity and constructs a 2.5-inch (6 cm) long cocoon around itself from fibers teased out of the surrounding plant material. The nymph metamorphoses into an imago in two to three weeks, but the juvenile can remain in the cocoon up to three months, until conditions are favorable for emergence.

The adults are good fliers but are not especially prone to leaving the place that hosted it. In fact, if a given palm has sufficient resources, several generations of this insect will live in it until the tree finally collapses into itself.

Nymph in its cocoon

Red Palm Weevil larva

AN INVISIBLE ATTACKER

Red Palm Weevil attacks are difficult to detect as the trees turn yellow and start to collapse just a few days before dying. It is important, therefore, to look for the signs: small perforations in the trees around the crown, the little bundles of fibrous material that have been ejected outside, and the noise made by the larva mandibles as they sever the core fibers.

Rhynchophorus ferrugineus

(Olivier, 1790)

Size: 40 mm

Rhyssa persuasoria

Saber Wasp

A huge "needle" to lay the eggs

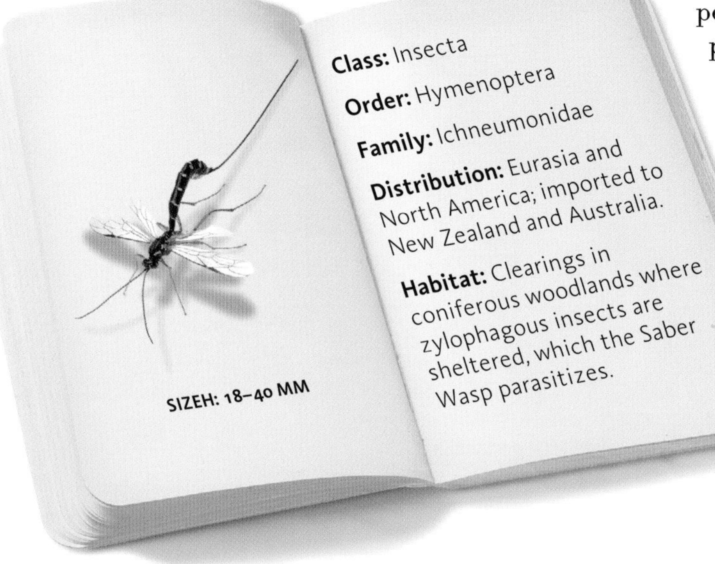

Origin of the name

The name originates from the Greek *rhusos*, for "wrinkled," and from the Latin *persuasoria*, for "that which persuades." Could it be that its long ovipositor (though completely harmless) persuades us to leave this insect alone?

Parasitoid wasps

The Saber Wasp is the largest member of the Ichneumonidae family of Europe. It can reach a length of 1.5 inches (4 cm) to which, at least in the female, we can add another inch and a half of length due to her incredibly long ovipositor. The Ichneumonidae are parasitoid wasps. Adults will fly, dip into in flowers and sip sugary liquids. Their larvae, however, can only develop to the detriment of another insect (or, more rarely, a spider), which supplies their food and that they kill in the process.

A long needle to "inject" her eggs

The Saber Wasp develops most often on the larva of another hymenopteran, the Giant Sirex (Urocerus gigas), which itself develops in weakened or freshly dead conifers. At egg-laying time, the female raises her abdomen in order place her long ovipositor perpendicular to the wood that she plans to bore into (opposite photo). This ovipositor is composed of four solid and supple pieces; two of these serve as protection and as guides to the two others that are true egg-laying organs. The ovipositor pierces the wood and allows the passage of the egg up to the Sirex larva (photo below). The female takes 20 to 40 minutes to drill into the wood and reach her victim. The egg will hatch and discharge a larva that develops into an ectoparasitoid – that is to say, on its host – which is now imprisoned in the tunnel it dug before becoming parasitized.

The Saber Wasp larva consumes the Sirex larva through several bites, being careful not to destroy the prey's vital organs, however. This way the wasp larva is guaranteed a steady supply of "fresh food" throughout its development! After hibernating, the wasp larva becomes a nymph in a cocoon of silk and wood fibers in the following spring. The entire development cycle lasts one year. When it emerges in summer, the juvenile must still make its way to the outside of the host tree.

The Saber Wasp was voluntarily imported to New Zealand and Australia in the 1930s in an effort to limit the proliferation of the damaging Sirex, which had been accidentally imported there along with some infested European pine trees.

Two wonders for the price of one

The very fact that the female Saber Wasp can pierce over an inch of wood to reach a larva on which she is destined to lay her eggs is certainly a wonder. This, however, presupposes that she can locate the larva within the enormous mass of wood she faces. To help with this, the ovipositor antennae have several sensory receptors to detect chemical and physical stimuli. The female runs along the oozing resin, tapping here and there with her antennae in order to locate the inner tunnels. Once she has accomplished this, the ovipositor takes over, searching for chemical signals from the Sirex larva from its excrement or from fungi that develop in its tunnel.

Saber wasp in flight

The target: A Sirex larva

Rhyssa persuasoria

(Linnaeus, 1758)

Size:

18–40 mm

ROSALIA ALPINA

Rosalia Longicorn Beetle

Class: Insecta
Order: Coleoptera
Family: Cerambycidae
Distribution: From France and Spain to central Europe, to the Caucasus and Syria; gone from Scandinavia for a century.
Habitat: Forests of broadleaf trees, especially on old oak and ash trees and occasionally willows, nut trees and alders.

SIZE: 15–40 MM

Mating

Longicorn Beetles mate on tree trunks during the warmest daytime hours. The much-courted female is often the object of affection of several males. Once fertilized, she lays her tiny rice-grain-sized eggs into the wood of dying or dead trees by deeply inserting her long and flexible ovipositor in the crevices of the bark. Unless it becomes premature fodder for a passing woodpecker, the xylophagous larva remains inside the wood just an inch or so below the surface for at least three years.

Origin of the name

As you might guess, the genus name of this insect refers to the name of a woman, Rosalie – probably the wife of the first person to describe it, Jean-Guillaume Audinet-Serville (1775–1858), cofounder and first president of France's Société Entomologique. It is also known as Rosalie of the Alps, a reference to the Swiss Alps, where naturalist Johann Jakob Scheuchzer first discovered this insect in 1793.

The most beautiful coleopteran in Europe

It is almost unanimous: the Rosalia Longicorn Beetle is one Europe's most beautiful coleopterans. Its elytra are gorgeous gray-blue velvet in coloration with black, unevenly spread out markings. The rest of the body is also bluish, and its long antennae are ringed with tuffs of black hairs. The sexual dimorphism is clear. The males' antennae are twice as long as their body (the same length as the female's body), and its mandibles are serrated.

Rotten wood lover

Adults emerge from June to September. They fly during the late morning, circling the trunks of ailing trees in order to lick their oozing wounds. They can also be found in large numbers on the logs of freshly cut beech trees and less frequently on umbelliferous plants.

Rosalie, if you please...

The Rosalia Longicorn Beetle is fairly common locally in the mountains, but is more typically found in plains and meadows. In France, it is found mostly in the Pyrenees, the Alps and the Cevennes in Corsica, but also on the shores of the Loire river. The species is experiencing a loss in numbers, however, due to the decline of old beech groves – its favorite domain. It has adapted by the very fact that its larva can also develop on a number of broadleaf trees, including ash, lime, willow, nut trees, hazelnut trees, maples and so on.

The species is globally protected by Annex II of the Berne Convention and by decision of the 22-VI-1993 at the national level. The Swiss have also become actively engaged in its protection by putting a program in place in three regions (Val Verzasca, Prättingau and Jura). This program, which involves planting dead beech tree trunks, which are about 7 feet (2 m) in length and at least 10 inches (25 cm) in diameter, in an upright position in sunny locations, has led to rapid reestablishment of this beetle in Switzerland.

ROSALIA'S STAMP

At least 15 countries have been so inspired by the beauty of the Longicorn Beetle that they have put its image on their postage stamps. These countries include Hungary, Romania, Switzerland, the former Czechoslovakia, Germany, Bulgaria, Croatia, Macedonia, Poland, the Ukraine, Serbia-Montenegro, the former Yugoslavia, and also Cambodia, Somalia and Togo (where the species doesn't even exist!).

The Rosalia larva

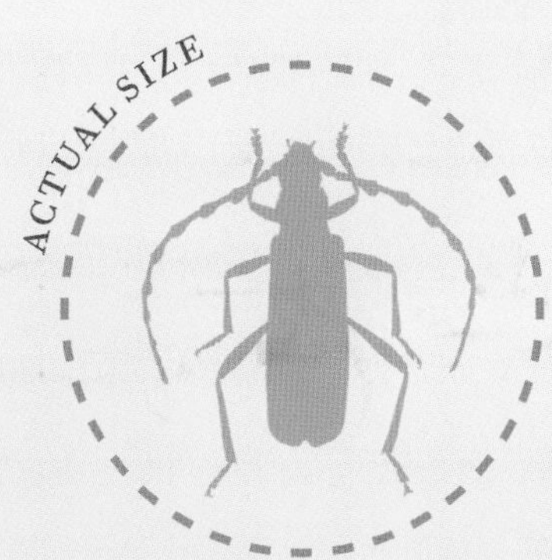

Rosalia alpina

(Linnaeus, 1758)

Size: 15–40 mm

Spotted Longhorn Beetle

Searching for pollen

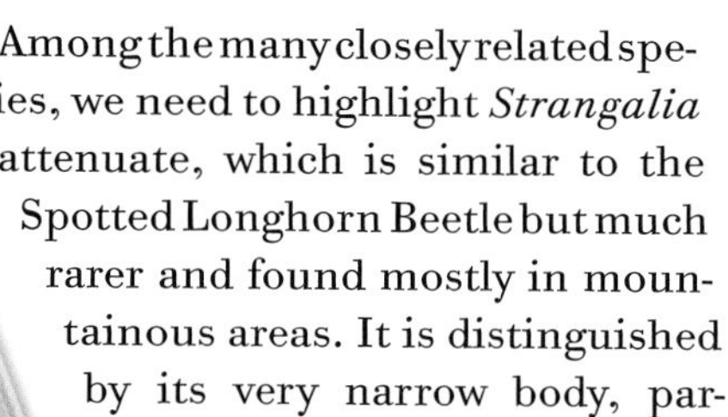

Among the many closely related species, we need to highlight *Strangalia* attenuate, which is similar to the Spotted Longhorn Beetle but much rarer and found mostly in mountainous areas. It is distinguished by its very narrow body, particularly narrowed toward the back, and by its elytra, which have four transverse black bands on a yellow background.

Class: Insecta
Order: Coleoptera
Family: Cerambycidae
Distribution: Europe up to the Caucasus and near-Orient.
Habitat: Prairies, clearings and forest edges; on umbelliferous or other flowers; the larva live in the dead wood of broadleaf trees.

SIZE: 13–20 MM

The Longhorn Beetle larva

An extremely common longicorn

As soon as the sun begins to warm hedges and road embankments near forested areas, the Spotted Longhorn Beetle perches on the pollen-rich flowers it enjoys feeding on. By flitting from one flower to the next, this insect is an effective pollinator. Bramble bush flowers, as well as those of white umbelliferous plants (*Achillea*, wild carrot and so on) are preferred, although it is also seen on many other corollas. The fearless Spotted Longhorn Beetle is widespread from April to September. With a diurnal habit, it is particularly abundant in July during the longest days of summer.

Origin of the name

The word "lepture" originates from the Greek *leptura*, which means "with a thin (*lepto*) tail (*oura*)," in reference to the thin, elongated extremity found on this insect's body; *maculata* means "spotted" (as in its elytra). Following the vagaries of scientific nomenclature, this species has been successively classified in both the genus *Leptura* (created by Linnaeus), and then *Strangalia* (with a "strangled" neck) – which explains why this insect is sometimes called the Spotted Strangalia – and, more recently (2003), *Rutpela*.

Secrets from an extraordinary acrobat

The Spotted Longhorn Beetle is easilty recognized by the bicolored segments on its antennae (a trait characteristic of the genus *Rutpela*) and by its black head and thorax, which contrast with the chamois-yellow-colored elytra, which are marked with symmetrical black spots. The body ornamentation is quite variable; for example, the elytra can be almost entirely yellow, with only their extremities being black or, inversely, the elytra can be entirely black.

The larvae in wood

During mating, which occurs on flowers, the male (full page image to the right) holds on to the female with a "tooth" found on the inside edge of his posterior tibia. Toward the end of his short life (only two to three weeks), the female lays about 100 eggs in rotten wood – such as the trunks of narrow felled dead trees or branches that have fallen to the ground – preferably in a damp environment. It almost always occurs in broadleaf trees (beech, oak, poplar, birch, and so on). Depending on the moisture content of the wood, the larvae will develop in two to three years. While in their little curved chamber (loge) from May to June, they then metamorphose into nymphs until the adults emerge.

WASPLIKE MARKINGS

Several insects that sip from flowers during the day sport "vespiforme" (imitating wasp ornamentation) markings in an effort to protect themselves against predators that are attuned to this warning. This explains why many hedge flowers are visited by all kinds of insects, including coleopterans, cipterans (syrphes) or butterflies, that display the same yellow and black coloration of a wasp. The Spotted Longhorn Beetle is one such insect. It carries the imitation even further by adopting the vigorous flight pattern, with low-hanging hindlegs, of predatory hymenopterans.

A male Spotted Longhorn Beetle

ACTUAL SIZE

Rutpela maculata

(Poda, 1761)

Size: 13–20 mm

Zebra Spider

SIZE: FEMALE 5–7 MM
MALE 5–6 MM

Class: Arachnida

Order: Araneae

Family: Salticidae

Distribution: Eurasia (although more frequent in northwestern Europe), North America and North Africa.

Habitat: On rocks and sunny fences, or in hot, dry meadows and in houses on windows.

The female Zebra Spider

ORIGIN OF THE NAME

Zebra Spiders derive their genus name from the Latin *saltus*, meaning "jump," in reference to the way they hunt by leaping on their prey. Their species name, *scenicus*, originates from the Greek, meaning "comedian" and also alludes to the costume-like markings on the abdomen of this small spider, which is harmless to humans.

SCALES TO PROTECT IT FROM THE SUN

The Zebra Spider, encountered in all areas of France, likes to live in houses. It seeks out sun and is thereby found on fences and walls around the outside of houses or, when living inside, on window panes. It can also be found in more natural settings sunning itself on sun-baked rocks.

The bodies of Zebra Spiders that live in dry, sunny environments are covered in scale-shaped silk deposits – the arthropod's sunblock! This insulating material, a unique development in the Zebra Spider, helps them avoid desiccation and dryness while minimizing water loss. This process is similar to insulating your attic to save on heating and cooling costs.

LOOK IN MY (EIGHT) EYES

With eight eyes, the Zebra Spider has very well-developed vision! Two pairs of eyes are located in front of the head, but the two middle eyes are particularly impressive. These "objective lenses," fringed with protective eyelashes, can fix their gaze precisely on prey, on incoming danger or on a sexual partner. The other two smaller pairs of eyes are located – one behind, the other on the side of the head – giving this spider a 360-degree field of view.

MINUS A WEB, BUT NOT MINUS A THREAD

Like all forms of spiders, Zebra Spiders secrete silk but do not, however, weave a web. Their silk is used to construct a retreat cocoon, located in a crevice where the spider seeks shelter while waiting for the sun to rise. This shelter offers protection during the particularly vulnerable time of molting and can also function as a nursery around the month of June. Although the mother lays a seemingly small clutch of about 40 eggs, the care she brings to them more than makes up for that small number.

While the Zebra Spider does not use silk in hunting, its prey-capturing method is nevertheless efficient and fearsome. It does not hesitate to attack prey as large as itself. It approaches its prey by scurrying around obstacles and jumping on it using its long muscular hindlegs. Each "jump" is accompanied by a secretion of silk that becomes anchored to the ground. At the moment of the jump, the exuded bit of silk is a bit slower than the speed of the jump so the spider can use the silk to control its positioning and equilibrium when landing back on its legs. It can also use the silk "rope" to escape from a bad situation very much like a prisoner using rope to drop down a prison wall.

Attacking large prey

A scary face!

ACTUAL SIZE
Salticus scenicus
(Clerck, 1758)
Size:
Female 5–7 mm
Male 5–6 mm

House Centipede

Juvenile House Centipede

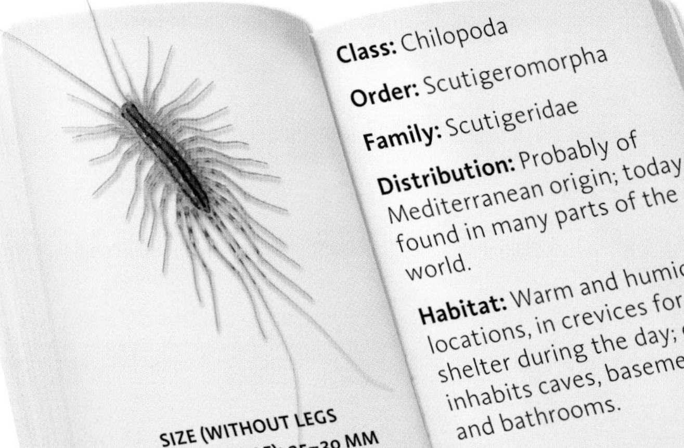

Origin of the name

The genus name *Scutigera* (from the Latin *scutus*, for "shield," and *gerere*, for "carrier") is based on this centipede's seven small plaques, striped with three dark bands, which cover and protect its back. One author saw similarities with spiders and gave it the species name *araneoides*; another called it *lineate*. Lastly, Linneaus saw a resemblance with a coleopteran and gave it the species name *coleoptrata*.

A poorly known ally

This centipede is hard to wrongly identify since only one species is found in France. The House Centipede likes humidity and heat. In Mediterranean regions it is found among cracks in rocks and caves. In cooler regions it has introduced itself into our homes. Cellars and bathrooms are its favorite haunts – not to everyone's liking! Nonetheless, the centipede helps rid us of woodlice (p. 186), flies (p. 180) and mosquitoes, as well as cockroaches (p. 50). Equipped with excellent vision (it is one of the few centipedes with large, multifaceted eyes like insects), and capable of remarkable speed, it chases its prey on walls once night sets in.

The House Centipede's rear and visibly longer pair of appendages are not used for locomotion – they are tactile organs. As these appendages lie in near-perfect symmetry with the centipede's long frontal antennae it seems like this animal has no head! The head is equipped with two very sharp mandibles and point out the centipede's carnivorous nature (photo at bottom left). Like its cousin the Brown Centipede (p. 160), the House Centipede kills its victims with the help of venom before eating it. However, it is not a concern for humans.

A waltz with a thousand legs

House Centipedes reproduce during the summer. Males and females circle one another with intertwined feet, as well as touch each other with their antennae. During this nuptial dance, the male deposits a semen-filled spermatophore on the ground. If needed, the male will guide the female so that she can grasp the spermatophore with specialized appendages (gonopods) and bring it to her genital opening. She will subsequently deposit between 100 and 300 isolated eggs into cracks and fissures. Larvae are born with four pairs of feet. They undergo six molts before attaining the final number of legs (15 pairs), and four additional molts before reaching adult size and sexual maturity. Adults live for two to three years.

An efficient predator. Note the large eyes and venomous pincers

THE BEST WAY TO WALK

The House Centipede moves its body "suspended" by its 15 pairs of feet. When resting, its body touches the ground. Its very long hindlegs often overlap with the shorter forelegs but never become entangled. Unlike the Diplopoda (like the Pill Millepede, p. 128), the paired legs of chilopods are not synchronized; when one set of legs advances, those on the opposite side of the segment stay on the ground. The House Centipede is extremely fast and can move up to 16 inches (40 cm) per second.

House Centipede on a wall

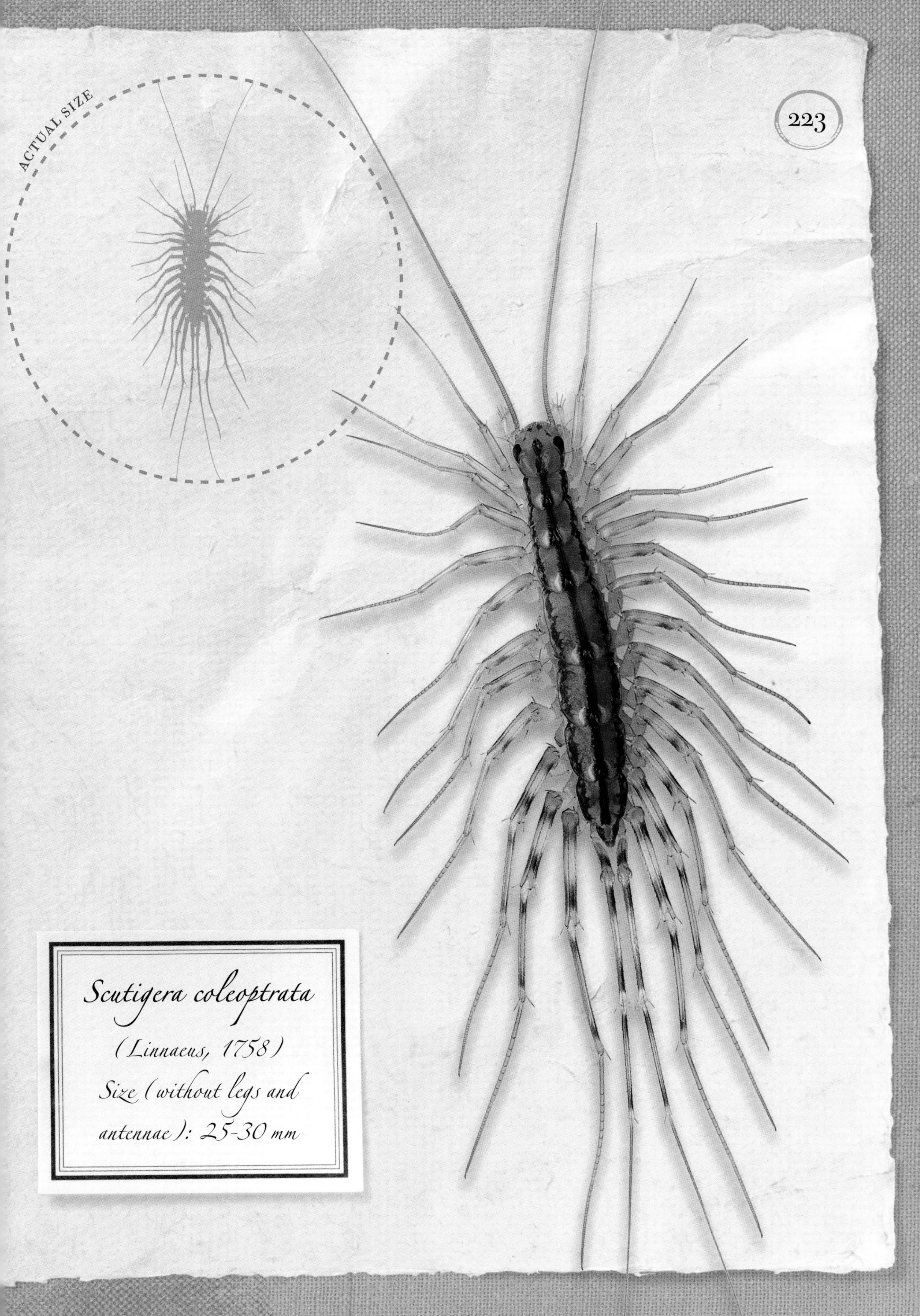
ACTUAL SIZE
Scutigera coleoptrata
(Linnaeus, 1758)
Size (without legs and
antennae): 25-30 mm

Giant House Spider

At rest in a bathtub

Class: Arachnida

Order: Araneae

Family: Agelenidae

Distribution: North America and Europe.

Habitat: Infrequently visited areas of buildings (cabins, barns, basements and so on); crevasses in dry rockeries.

BODY LENGTH:
FEMALE: UP TO 18 MM
MALE: 14 MM
(BUT WITH LONGER LEGS)

Origin of the Name

The genus name *Tegenaria* is derived from a Greek word meaning "roof" and, by extension, "house or habitation," in reference to the locations where these spiders are found. The species name *duellica*, "belligerent" in Latin, is hardly appropriate for a spider that flees at the first sight of danger. *T. gigantean* ("giant"), another accepted name, shows how difficult it is to distinguish different species of this genus. Their colors and markings are not reliable criteria.

Arachnophobia

If one of these spiders shoul suddenly appear under a cupboard, screaming is inevitable in many households. Arachnophobia (fear of spiders) is the most common fear among humans, and in our region is most often caused by the Giant House Spider. Since they don't climb smooth surfaces very well, it's not uncommon to find an impressive specimen in the bathtub. However, this spider is timid and completely harmless. So overcome your fears, cover it with a glass and slip a piece of paper under it. That's all this spider wants from you!

A Very Useful Spider

With some 70 to 90 species in Europe alone, spiders of the genus *Tegenaria* are really part of our landscape. While some live "with us," others live in forests and countryside areas. All feed on flies and mosquitoes and in that sense are very useful. All these spiders spin triangular-shaped webs; these end in a mesh of tube-shaped silk, and are usually located in the corner of a garage, or in sheltered areas of the house and garden. Outside, they generally establish themselves in hollow trees, in caves or amid very dense vegetation. Hidden like this in its retreat, the spider waits a fly to become trapped in its web. Patient and hardy, it can survive for months without prey and water.

The Female Stays at Home, the Male Wanders

In the fall (sometimes in the spring), males wander in search of a female to mate with. The female, which lives for several years, is quite sedentary. To keep it simple: the big spider in a web located deep inside your garage is a female; the spider crawling around your bathroom is a male. After mating, the male stays with the female for several weeks before he dies. There is little doubt she feeds on the remains of her companion. The female, meanwhile, spins a white silken cocoon and deposits her eggs in it. The juveniles spend part of their development there, then live alone once they're on their own.

NO FIXED ADDRESS?

The Hobo Spider, as it's sometimes called, is none other than our field *Tegenaria* (*T. aggrestis*), which was accidentally introduced to North America in the 1930s. Living naturally in Europe, it readily adapted to the New World, where it has no competitors. An unusual fact is that it has become aggressive in North America and its bite can cause nausea and tissue necrosis, while in Europe such behavior has never been reported in this species. Some scientists remain dubious, however.

Catching a fly

House Spider cocoon

ACTUAL SIZE

Tegenaria duellica

Simon, 1875 (= *T. gigantea*
Chamberlin (& Ivie)
Body length: Female: up to 18 mm
Male: approx. 14mm

Mealworm Beetle

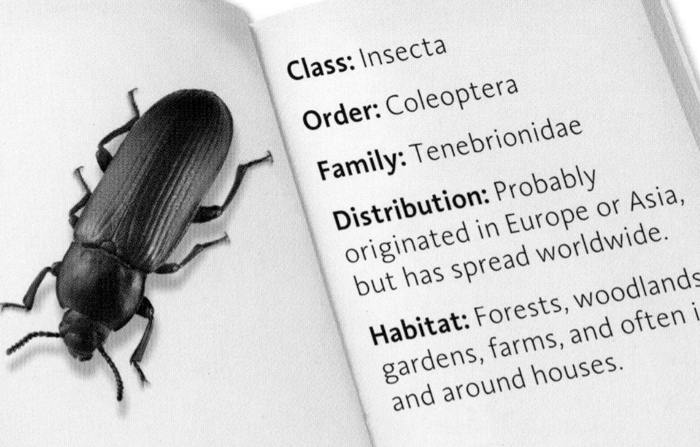

An emerging adult, not yet pigmented

ORIGIN OF THE NAME

The genus name *Tenebrio* is derived from *tenebra*, or "obscurity," a reference to the Mealwrom Beetle's somber coloration and to the fact that it frequents obscure areas. The species name *monitor* means "miller," because the larvae develop in a variety of detritus and, when it can, in pastry flour (its French name is Flour Beetle). In days gone by, it was not unusual to discover one of these little beasts actually baked in a loaf of bread.

A DIET OF BREAD AND WATER

Like the majority of Tenebrionidae, the Mealworm Beetle is adapted to a rather dry environment (most species are found in desert-like habitats). This insect has the unique ability to not drink water but rather recover it from the atmosphere or from the little amount left in dry matter. It is safe to assume that, in our region, this beetle originated in the nests of various animals, since today it is found in our dwellings where its larva feeds on wheat flour, bread, bran muffins, pastries, dried fruit, leather, cured meats and so on. It is also found in chicken coops, where it feeds on feathers and dead insects.

A RESILIENT INSECT

The Mealworm Beetle has been seen to go without water for long periods of time. Its long larva, with brown-yellow "varnished" skin (the infamous "flour maggot"), is also very resistant to heat, cold and fasting. Its growth can be fast or slow, and it goes through seven to 20 larval stages, depending on the amount of food available. The resulting adults, therefore, also vary in size.

VERY EASY REARING

Adults live just 50 short days, and though they are equipped with functional wings, hardly ever fly. The female will lay about 30 eggs a day, up to 200 to 300 eggs. The larval phase can last three months under optimal growing conditions (77°F–81°F/25°C–27°C) and up to a year in the wild.

Tenebrio larvae are easy to find in bait stores, where they are sold to fishermen, and as in pet stores as food for birds and reptiles. Larva-rearing is so easy even a child could do it, especially since ordinary cat food can serve as nourishment for these little wormlike creatures. Mealworm Beetles are also very commonly used in research laboratories as "guinea pigs" in physiology, genetics, or to do chemical assays (such as with pesticides).

ENTOMOPHAGY

Mealworm Beetle larvae (flour maggots) constitute a gourmet meal – once cooked – for those of us who practice entomophagy (an insect-based diet). They apparently taste quite good. For example, at the Insectarium in Montreal, Quebec, it's possible to purchase delicious Mealworm Beetle lollipops! It is important to avoid eating them raw, however, since they can carry diseases or parasites. For the very same reasons, humans should abstain from eating the larvae sold in pet stores.

Mealworm Beetle larva (above) and nymph (below)

ACTUAL SIZE

Tenebrio molitor

Linnaeus, 1758

Size:

11–24 mm

Great Green Cricket

Great Green Cricket larva

ORIGIN OF THE NAME

This cricket is also known as a Katydid. The genus name *Tettigonia*, drawn from the roman naturalist, Pliny the Elder (1st Century A.D.) means "small cicada." The species name, *viridissima* ("very green"), obviously refers to its color.

BIG WINGS AND LONG ANTENNAE

This splendid cricket is distinguished by its large size and its lustrous green coloration. Sometimes their long antennae have been invoked as a distinguishing feature from those of the crickets and grasshoppers (see p. 76 and p. 136, respectively); this, in some classifications, place them in a separate suborder. Their long wings and elytra overflow out from their abdomen. The Great Green Cricket flies very well and can cover a distance of 65 to 130 feet (50 to 100 m) before landing. Its well-developed hindlegs enable it to jump very well. When it is not chirping, it can easily pass unnoticed because it is well camouflaged amid vegetation. Once quite common in farmlands, it has since become the victim of agricultural equipment.

THE LEGS HAVE EARS

Positioned in trees or bushes, the males chant from night until daybreak. Its powerful chant is reminiscent of clanging cymbals. The "singing" attracts females, which are equipped with an auditory organ located under the knee joint of their hindlegs. The males are also equipped with a similar organ, though they use theirs to locate rivals. This leads to singing "duels" between males; the left elytron has a a row of minuscule teeth (the bow) that rubs against a membranous segment on the right elytron (the eardrum). The chant frequency increases with the rise in ambient temperature.

MATING BY PROXY

During mating, the male deposits a spermatophore (a semen-laden pouch) against the female's genital opening, and the transfer of spermatozoa occurs (this process takes from 30 to 45 minutes). The female devours the empty spermatophore. She lays about 100 eggs in small packets in the ground with the help of her sharp, saber-shaped ovipositor. The young hatch the following year, during the second spring of their life, and the juveniles feed first on aphids and then on small flies, caterpillars and other insects, as well as low-growing plants. Before becoming adults, the juvenile must go through two larval stages.

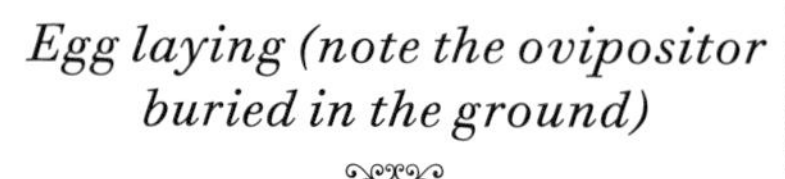

Egg laying (note the ovipositor buried in the ground)

Male Cymbal Cricket
(Tettigonia cantans)

LITTLE KNOWN GARDEN COMPANION

The Great Green Cricket is a carnivore that eats very little vegetation as compared to the huge amount of ingested insects – various caterpillars and larvae, including those of the potato beetle are devoured in large quantities, crushed by the insect's powerful mandibles. Keep this in mind if you ever capture one of these insects: they can easily and quite forcibly pinch your fingers.

Tettigonia
viridissima
(Linnaeus, 1758)
Size:
Male: 28-36 mm
Female: 32-42 mm
ACTUAL SIZE

THAUMETOPOEA PITYOCAMPA

Pine Processionary Moth

Class: Insecta

Order: Lepidoptera

Family: Notodontidae

Répartition: Southern Europe, North Africa, Near Orient. Increasing in Europe due to climate change.

Habitat: On pine trees (especially the Austrian Black Pine, Sylvestre and Maritime Pines).

WINGSPAN
MALE: 25 MM,
FEMALE: 30–35 MM

Processionary caterpillar and chrysalis

ORIGIN OF THE NAME

The name of the genus is derived from the Greek *thaumatos*, for "miraculous" or "object of awe." The unique manner in which these gregarious insects move about justifies their moniker, including their common name "processionary." *Pityocampa* means "caterpillar (*campa*) of pine (*pitus*) trees" in Greek. The ancient name of the genus, *Cnethocampa*, for "caterpillar that causes skin itching" in Greek is quite self-explanatory.

PINE TREES WRAPPED IN SILK FOR THE WINTER

The life cycle of the Pine Processionary Moth lasts about a year. The eggs, laid in groups of 100 to 200, are covered in scales originating in the female's abdomen. They actually form a sleeve around the pine needles. After hatching, the young caterpillars build a collective cocoon of silk and leave it every night in search of pine needles. The silk that is secreted during these nocturnal missions constitute what is known as "Ariane's thread," which the caterpillars follow as a guide to return to the cocoon. During winter, this material is used to insulate a much more complex structure that acts as both a refuge and a heated confine.

Predator-proofed collect cocoon

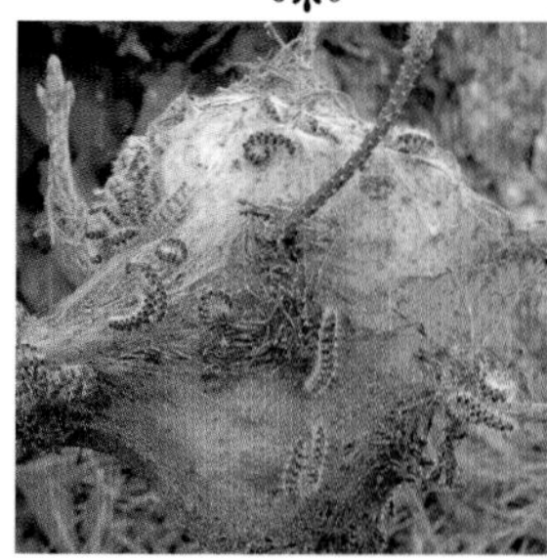

MARCHING TAIL-TO-TAIL

The adult Pine Processionary Moth's faded appearance and nocturnal behavior allows it to move around almost unnoticed. However, its caterpillar is legendary and feared. In spring near pine trees, it is not unusual to see dozens or even hundreds of these little caterpillars scurrying along tail-to-tail in single file in a lineup that can sometimes be more than 30 feet (9 m) long! Each caterpillar is in contact with the one ahead of it and secretes a thin silk thread to maintain the unity of the line. By winter's end and in spring, these spectacular processions lead the caterpillars to a favorable site where they will be able to metamorphose. While underground, the caterpillars weave a silk cocoon around themselves in which they can transform into moths. Adults emerge at the end of summer. Following fertilization, the female immediately begins her search for the ideal conifer – usually a pine or, more rarely, a cedar – where she will lay her eggs.

PINE TREE PESTS

The Pine Processionary Moth is one of the most devastating forest insects, as conifers become greatly weakened after their needles are devoured by these larvae. In addition, when the caterpillar is threatened, it releases irritating hairs, which are capable of causing serious skin allergies and lung reactions in humans, as well as in some domestic animals such as dogs. Pine Processionary Moths nests, even when empty, still retain their allergenic properties

THE STRUGGLE AGAINST PROCESSIONARY MOTHS

Even though its silk has irritating properties, this caterpillar does have predators, such as the gray cuckoo and the titmouse, which approach the caterpillars' nests and grab them on the fly. Some dipterans and ephippigerans (related to grasshoppers) are also good predators. Southern Brown Wood Ant (p. 124) nests have been disturbed during these struggles against the moth. Another widespread means of controlling them is the use of biological insecticides (such as those based on *Bacillus thuringiensis*).

A procession of Processionaries

Thaumetopoea pityocampa

(Denis & Schiffermüller, 1775)

Wingspan: Male: 25 mm

Female: 30-35 mm

Timarcha tenebricosa

Bloody-nosed Beetle

Blood-spitter of the species Timarcha goettingensis

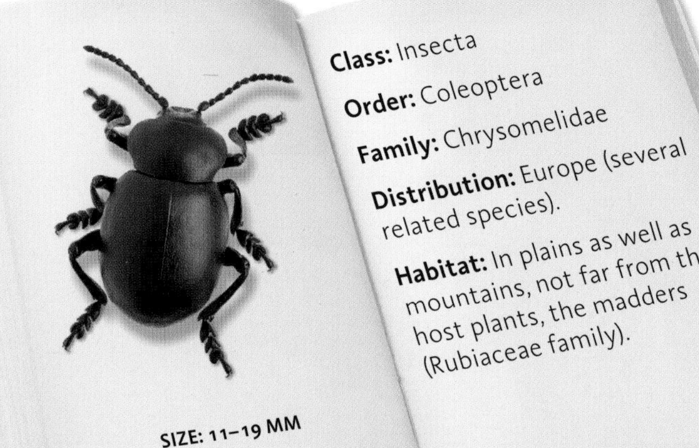

Origin of the name

The name *Timarcha* refers first to the voters of Athens and then the Roman censors, both elite members in antiquity who needed to demonstrate their status by a disguised and phlegmatic manner of walking – a nod to the slow and measured walking pattern of these insects. Their black color has given them the species name of *tenebricosa* ("dark"). This species earns its common name from its peculiar defense mechanism: when threatened it exudes a drop of bright red fluid from its mouth (the fluid originates from articulations in its legs).

A beetle raised high on its legs

With its rounded, massive body, this big oaf is reminiscent of the scarab beetle. Its long legs, ending in tarsi with four articulations, hints strongly at its similarity to the Chrysomelidae. The Bloody-nosed Beetle is found in sunny locations, including roadside hedges, on madders (notably *Gallium mollugo*) on which its large, black and chunky larva (up to 20 mm long) feed. The male, smaller than the female, has enlarged anterior tarsi to better grasp on to the female during mating.

Differed Mating

Adults emerge in July or at the beginning of August. Apart from a few matings followed by egg laying in September and October, the majority of adult Bloody-nosed Beetles will hibernate in their nests or in a shelter without first reproducing. Activities resume at the beginning of March. The female lays about 100 eggs and dies, as do most adults, at the end of spring. The eggs that were laid in spring and those, less numerous, that are laid at the end of summer by newly emerged adults, hibernate for one winter before hatching (the complete cycle spans one or two years). The larva goes through three stages and becomes a nymph while sheltered underground. So basically, each fall we find that year's adults, the eggs from the second brood, and the larvae issued from hibernating springtime broods. Since the eggs from the second brood do not hatch until February to March of the following year, there is really only one generation per year.

Bloody hit

When threatened, the Bloody-nosed Beetle plays dead. If the interloper persists, the insect emits a foul-tasting bloody red liquid from its nose and from articulations in its legs. This phenomenon is known as "reflex bleeding" or, more scientifically, autohemorrhaging. This reflex secretion is found in other insects, too, notably the Ladybug (p. 88) and the Violet Blister Beetle (p. 172).

Larva on its host plant, the madder

Mating: the male hangs on to the female with the enlarged tarsi on his hindlegs

UNABLE TO FLY

The Bloody-nosed Beetle cannot fly because its elytra are fused. It would be easy to see this as a handicap, except that fused elytra provide better protection for this insect (for example, against dehydration). In addition, the space freed by the reduced wing musculature provides more room for larger eggs, which are richer in egg yolk and resistant to cold temperatures. Another benefit of this anatomical peculiarity is that the inability to fly conserves energy which then may be tapped into during reproduction.

Timarcha
tenebricosa
(Fabricius, 1775)
Size:
11–19 mm
ACTUAL SIZE

TIPULA OLERACEA

Marsh Crane Fly

Marsh Crane Fly larva

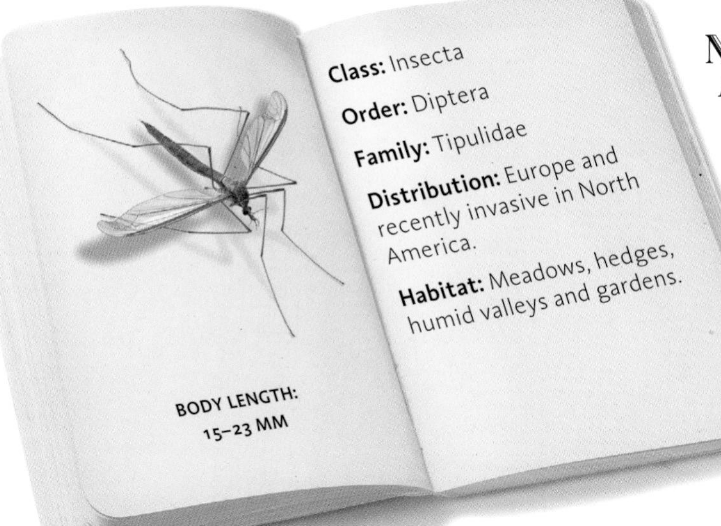

Natural history

Adults emerge from the ground at spring. Mating often occurs at night. Males seek out females in vegetation, fly around in a dancelike manner and then drop to the ground. The female lays her eggs under leguminous plants such as alfalfa and clover, as well as fields of cabbage, sheltered by the plants' leaves. The method of egg laying is surprising: she flies an inch or two above the ground and drops down on her hindlegs. She then inserts her ovipositor into the soil and forcibly ejects a "salve" of small black eggs. She repeats this process several times.

Front-to-back mating

Origin of the name

The Marsh Crane Fly is also called Cabbage Crane Fly, Harvestman Fly or Cousin. Its long, spindly legs and its attraction to humid environments have also given this insect the name of water spider, in Latin *Tipula*. Its species name, *oleracea*, which means "vegetable garden" in Latin, informs us that this insect's larva is a common find in farmer's markets.

Large, fake mosquitoes

Marsh Crane Flies are dipterans in the suborder Nematocera (characterized by thin, threadlike antennae) with an elongated head in the form of a muzzle. Despite their fearsome appearance, these are not mosquitoes; plus, these flying cousins don't even bite! Like all dipterans, Marsh Crane Flies only have one pair of wings. The second pair of very visible wings are modified flight stabilizers.

Big worm for such a slim adult

The gray-green larva grows quickly to a length of 1.2 inches (3 cm). It breathes through orifices located at the extremity of its abdomen, which are surrounded by pointed expansions (the form and position of these are characteristic of each species, aiding in identification). Lacking a visible head and having yet to develop legs, the larva uses the fine silk protrusions from its lower surface to crawl. It feeds in soil on fresh plant roots – in large numbers it can be quite damaging to crops. When it rains or if the humidity elevates, the larvae rise to the surface to feed on vegetables.

The nymph stage occurs at the end of the summer. When ready to emerge, the nymph undergoes a series of contractions and, with the aid of spines, crawls to the surface and emerges as a young adult. Cold weather forces the adults into our homes, but remember that they are completely harmless.

RECOGNIZING THE MISTER AND THE MISSUS

Since sexual dimorphism is well established in this species, it is therefore quite easy to distinguish the male from the female. The abdomen of the female, with its ovipositor, ends in a point. On the other hand, the male's abdomen (opposite photo) is squarer and terminates with its reproductive organs. Moreover, males have noticeably longer antennae.

The abdomen in the male

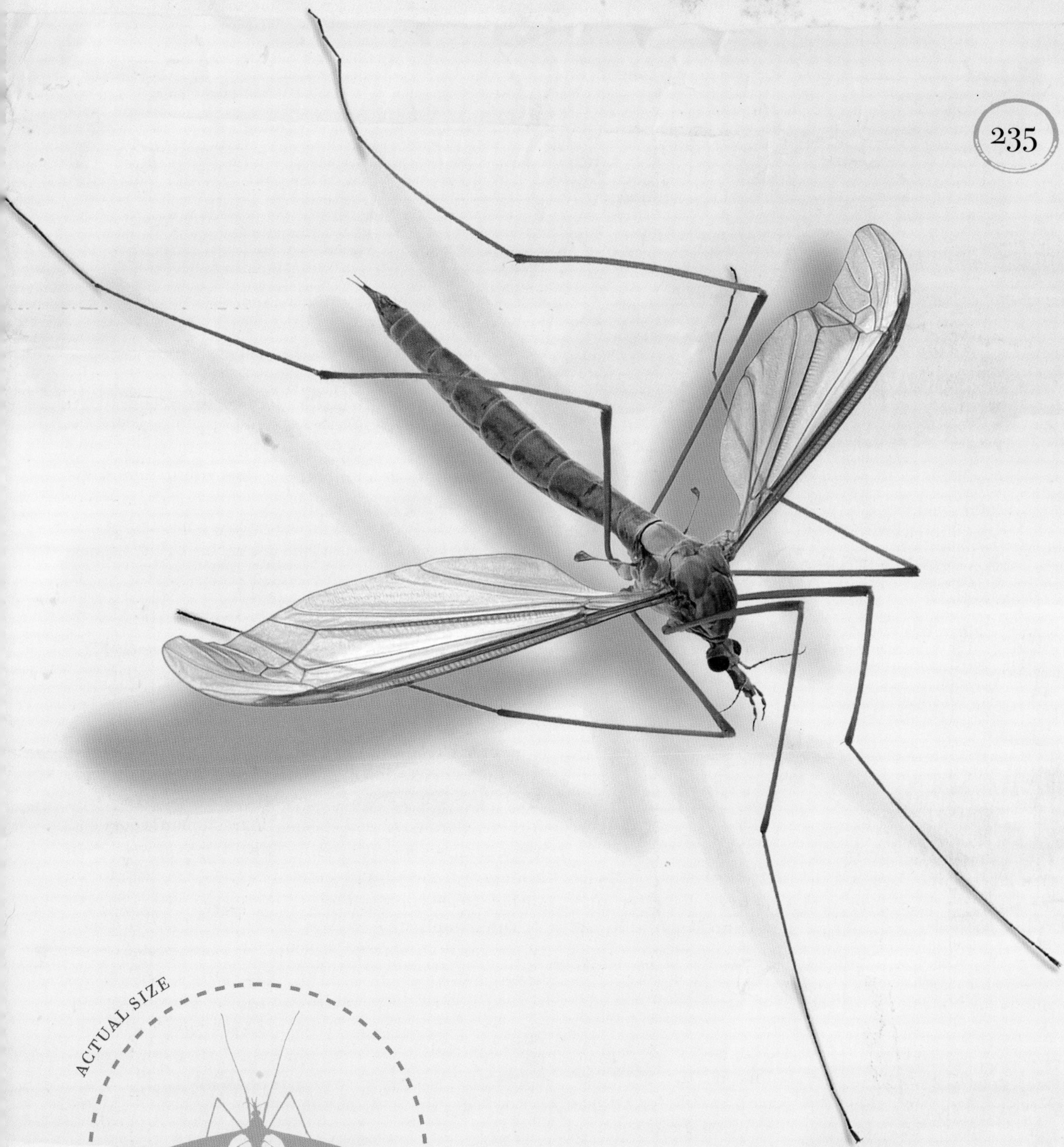

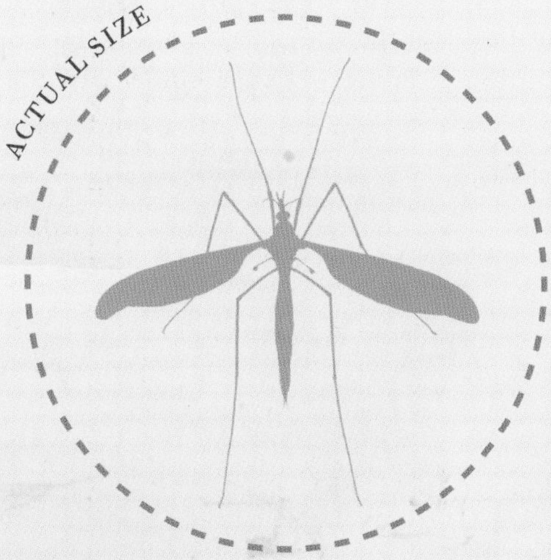

ACTUAL SIZE

Tipula oleracea

Linnaeus, 1758

Body length: 15–23 mm

TRICHIUS ROSACEUS

Rose Beetle

Trichius fasciatus

A TRUE FLOWER LOVER

Rose beetles are found on flowers of the Rosaceae family (rose and bramble bushes), but also on umbelliferous flowers, elder trees, spireas and daisies. We see it flying around in the daytime at the end of spring and early summer. Its furry texture and light brown elytra marked with dark spots resemble a bumblebee. Rose Beetles feed on flowers and specifically pollen. Moving from flower to flower, and transporting pollen via its "furry" coat, it also helps with pollination. As with many flower-loving species, mating also takes place on flowers. Eggs are laid mainly on birch, elms and decaying beech trees. The larvae live in old tree stumps where they feed on rotting wood. They mature in about two years and build themselves small, also in wood, to become nymphs.

ORIGIN OF THE NAME

The genus name *Trichius* is derived from the Greek *trichos*, for "hairs," in reference to the fuzzy head and lower face of this insect; *rosaceus* refers to its preferred flowers, those of the Rosaceae family. This pretty insect is also known as the Common Rose Beetle, as well as the Zoned Beetle because it has also been described under the name *T. zonatus*. When Geoffroy described this insect, he tagged it with the odd name of "Ancre," because the Marquis d'Ancre dressed his valets with a yellow uniform bordered by a yellow-green stripe!

A MIX OF THREE ROSE BEETLES

Rose Beetles are related to the Rose Chafer (p. 74), and even closer to the Riffle Beetle (p. 190). Three very difficult to distinguish species are found in France. The more common and widespread Rose Beetle has two very distinct black markings at the base of its elytra (near the thorax). In the more mountainous Bee Beetle (*T. fasciatus*), the two spots are more or less linked into a single band (see photo above). Lastly, the least common species, *T. sexualis*, is confined to eastern regions (Vosges and Alsace). In males, the posterior half of the abdomen is whitish, while the abdomen of males of the other two species has one or more smaller white spots. In any case, the positioning of spots on the elytras is so highly variable, especially among *Trichius rosaceus* and *T. fasciatus*, that it becomes difficult to determine which species it is. Experts can differentiate them, however, through the reproductive organs of each individual species.

Rose Beetle gorging on pollen

HOW TO DIGEST WOOD

The Rose Beetle is xylophagous – that is, it nourishes itself on decomposing wood. It could not digest cellulose (the main component of plant cells), however, without symbiotic microorganisms (protozoans and bacteria) that live in its gut. In reality, therefore, the beetles are nourished by the breakdown products of cellulose produced by the microorganisms. Among xylophagous insects, only a few, like the Silverfish (*Ctenolepisma lineate*), are capable of directly digesting cellulose.

ACTUAL SIZE

Trichius rosaceus

(Voet, 1769)

Size: 9–13 mm

Minotaur Beetle

Female Minotaur Beetle

This "coworker" of the Dung Beetle (p. 32) has developed a complex system of nest building.

Laying eggs at great depth

After surfacing in the fall, the female begins by digging a small shelter in sandy terrain. The male joins her there in the spring to build an enormous nursery. This takes them several weeks. The shelter can have several openings and extends to huge depths (up to 5 feet/1.5 m – a record among European dung beetles!), and at the deepest point branches into four to 15 secondary tunnels, each ending in a nesting chamber.

The male leaves to look for dung droppings after the first chamber is finished, while the female lays a single egg in each chamber and blocks the secondary tunnels with dung "provisions."

Origin of the name

The genus and species names, which are almost identical except for one letter, evoke the name of the giant Typhaeus, the evil deity that was buried under Mount Etna after being brought down with considerable difficulty by Zeus himself. The Minotaur, a monster with a man's body and the head of a white bull, was imprisoned in the labyrinth built by Daedalus and could only be killed by Theseus. Like our insect, the Minotaur was a horned being living underground. But what a terrifying name for such a peaceful insect! As Jean Henri Fabre so eloquently wrote in his *Souvenirs Entomologiques*, "The mythological Typhaeus wanted to destroy the realm of the gods by erecting a battery of severed mountains at its base. The Typhaeus of nature does not climb, however, but descends and pierces the soil to great depths."

400 rabbit droppings per nest!

The work required in this undertaking represents a colossal effort on the part of both sexes. Each task is very distinct; as the female works the soil, the male moves it out with its thoracic "trident." Sheep or rabbit droppings are pushed with the head or hindlegs to the tunnel opening and then broken up inside the nest. A single nest may require up to 400 rabbit droppings! The parents die after the work is finished. At that point, the nest contains larvae at various stages of development, which lasts from five to six months. Adults emerge in the fall.

An expert on rabbit dung

The male of this large black member of the Geotrupidae is armed with three spectacular forward-pointing horns on its thorax. The female does not have the lateral horns, just two small protrusions. The Minotaur Beetle is found in the woods and forests, and specializes in the dung of small herbivores (sheep, rabbits and deer).

Minotaur "sniffing" the air with its antennae

MINOTAURS VS. SCARABS

Even though they resemble scarab beetles, the Geotrupidae are a distinct family. Their legs have two to three articulations, while scarabs always have two. Furthermore, adults have antennae with 11 segments and exposed mandibles, while Scarabidae antennae consist of seven to 10 segments and their mandibles are covered.

Typhaeus typhoeus
(Linnaeus, 1758)
Size: 15–22 mm
ACTUAL SIZE

Red Admiral Butterfly

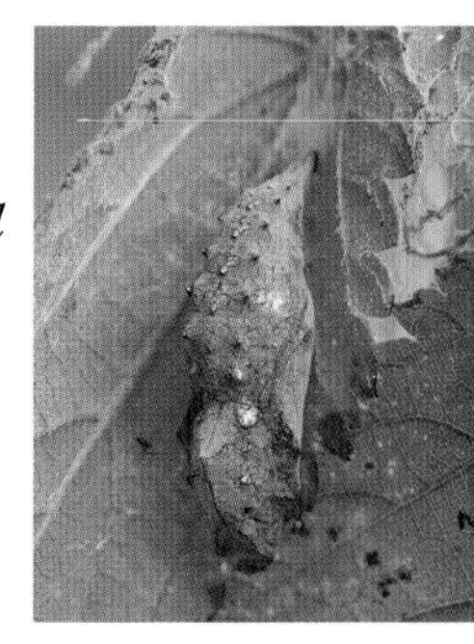
The Red Admiral chrysalis

Class: Insecta
Order: Lepidoptera
Family: Nymphalidae
Distribution: Temperate zones of Eurasia and North America, North Africa; introduced to New Zealand.
Habitat: Meadows, hedges humid valleys and gardens.

WINGSPAN: 56–63 MM

The Red Admiral caterpillar

Origin of the name

Fabricius coined the genus name *Vanessa*, which already existed in antiquity (it is derived from the Greek *phainein*, meaning "brilliant"). The species name *atalanta* is the name of a mythological female hunting figure, Atalanta, sometimes comparable to Artemis. The common French name of Vulcain refers to the Greek deity of ironsmiths, possibly named because of the red "flames" that mark the upper surface of this butterfly's hindwings. In English, we choose to identify more with the "costume" aspect of the wings and commonly refer to it as the Red Admiral Butterfly.

A four-footed butterfly

This magnificent diurnal butterfly beautifies our hedges and gardens from March to November. It is very active in the sun and flies with surprising speed; perhaps the warmth of the sun helps to somehow accelerate this butterfly. It is one of the most common butterflies in our region, making it that easy to observe. One of its traits, common to all Nymphalidae, is that its anterior legs are atrophied and covered with long brush-like hairs. Stuck to the face, these legs are almost invisible, making us think that this insect only has four, rather than a total of six legs.

A regular migratory species

The widespread nature of this butterfly is closely linked to its ability to migrate over long distances. Originating in warm regions, the Red Admiral Butterfly migrates north each year, although in diminishing numbers at polar extremities. Some years, it is common to them venturing deep into cities and drinking from flowers, especially those of the Butterfly Bush and Ground Ivy. In the fall, some individuals undertake a new southbound migration although, we now see (perhaps due to global warming) more adults hibernating locally and emerging when the weather warms up.

Lonely caterpillars

The female lays her eggs at the very edge of sun-drenched pellitories and nettle leaves. After they hatch, the caterpillars both feed on and seek shelter in these very leaves. Unlike many other caterpillars, the Red Admiral is not gregarious. Transformation into a chrysalis occurs on the host plant. Adults will need to spend time feeding to build up reserves for their long migratory flights. In addition to the flower's nectar, they also seek out the dripping tree sap and the juice of mature orchard fruit trees (prunes, pears and apples) and in mature hedges. The Red Admiral Butterfly readily lands in trees.

Detail of the scales

At rest – count its legs

MASS MIGRATIONS

Finnish ornithologists have observed the fall migrations of the Red Admiral Butterfly in their country. In the best year (1998), during the month of September, 1,240 migrating individuals were counted from atop a bird-observing station. From this, we estimate that about a half-million butterflies were migrating along a 60-mile (100 km) front. Radar tracking revealed that they flew at very high altitudes (out of the field of view) – a fact that may contribute to their presence in regions where they have very little chance of surviving.

ACTUAL SIZE

N.° Vanessa atalanta

(Linnaeus, 1758)

Wingspan:

56–63 mm

Vespa crabro

European Hornet

Origin of the name

The names *Vespa* and *crabro*, both originally from the Latin mean, respectively, "wasp" and "frelon" (this insect's name in French), which comes from the Frankish *hurslo*, via the Latin *furlone*. This large stinging hornet is distinguished from the "small wasps" (*Vespula* species; see p. 244). The European Hornet has piqued the popular imagination and has resulted in many common names, as well as it simply being called the Hornet.

A big fly destroyer

Apart from its size, the European Hornet is further distinguished from the wasp (p. 244) by the ferrous-red hue on its thorax. Hornets have the ability to capture all kinds of insects, though they prefer flies. They also attack bees by penetrating the hives to steal the honey. After killing their prey, they decapitate them and eat their insides, but only partially, saving the rest for the larvae and the queen. They also like to eat ripe fruit such as pears and prunes. On rare occasions, they will lick the sap of wounded trees.

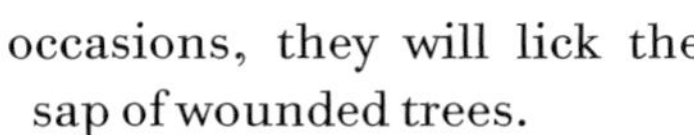

Workers in compartments

Once done with hibernation at springtime, the queen European Hornet makes a nest in a hollowed tree, a hole in the wall, underground or under a roof overhang. The hornet's nest is formed by layers of "cardboard," which consists of wood shavings that have been chewed and mixed with saliva. The queen lays one egg in each cell within the nest. After four to five weeks, the first set of workers hatch and assist the queen in building compartments and supplying nutrients to each cell. At its peak, the colony can consist of a few hundred individuals. By the end of summer, the future queens and males emerge and begin mating. A special substance known as royal jelly is kept only for fertilized females, which conserve the semen from several males in a special reservoir (the spermatothecum). These females hibernate under tree bark, in splits in wood planks or above ground.

Are they really dangerous?

The European Hornet is a peaceful insect and very useful in controlling fly populations. It does not attack humans, except when we disturb its nest. Its sting is quite painful (and can be felt for a period of about 48 hours), but several stings would have to be inflicted before a human life would be endangered. Obviously, people who are allergic to the sting are the most endangered, though not more so than those allergic to bees or wasps.

Serial Bee Killer

Recently discovered in France and now widespread in 18 southern regions, the Asian Hornet (*Vespa velutina*) originated in Southeast Asia. This is a fearsome enemy of apiaries and bee cultivators. Their thick skin protects them from bee stings and allows them to conduct "commando style" attacks and kill up to 30 percent of bees before pilfering the larvae's reserves. In their place of origin, bees are known to meet an untimely death, either through the heat of thousands of hornets pressed against them or by suffocation from their numbers.

Hornet enjoying overripe fruit

The paper-like texture of the hornet's nest

Vespa crabro

Linnaeus, 1758
Size: Worker: 18–25 mm
Male: 21–28 mm
Queen: 24–35 mm

Common Wasp

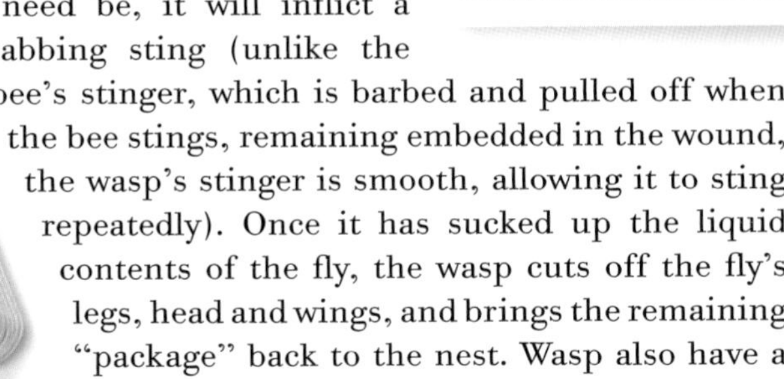
A true wasp's nest

The wasp mainly hunts insects, notably flies. It approaches the posed fly, grasps it with its legs and kills it with its mandibles. If need be, it will inflict a stabbing sting (unlike the bee's stinger, which is barbed and pulled off when the bee stings, remaining embedded in the wound, the wasp's stinger is smooth, allowing it to sting repeatedly). Once it has sucked up the liquid contents of the fly, the wasp cuts off the fly's legs, head and wings, and brings the remaining "package" back to the nest. Wasp also have a "sweet tooth" and often feed on sugary substances, such as flower nectar, ripe fruit and, in households, honey, fruit juice and jams.

Origin of the name

The Latin word for the Common Wasp, *Vespa*, is today the genus name of the European Hornet (p. 242). The true wasp has been given the name of *Vespula*, which is a diminutive form of *Vespa*. The name of the species, *vulgaris*, simply means "common."

Predator and gourmand

The Common Wasp is difficult to distinguish from the other wasps in Europe, such as the German Wasp (*V. germanica*), another widespread species. The Common Wasp can be seen in meadows, grassy hills, in clear woodlands or pine forests, at rivers' edges and on roadsides. It also ventures close to houses.

Wasps also like jam

PAPER PASTE

Wasps construct nests of "cardboard" made of wood. The wood shavings are chewed and mixed with their saliva to make a paper paste that is laid down in successive layers. In Réaumur's days, remarkable precursor of modern entomologists, paper was only made of chiffon. While observing the work of wasps in his Charenton property, he proposed that we find inspiration from these insects to address the approaching shortage of paper. A full century went by before we came to realize that wasps knew how to make paper for a very long time.

Well-organized insects

The queen awakens in the spring and builds her nest in a small subterranean hole with some passageways. She then lays her first eggs and, four to five weeks later, begins to rear the first generation of workers. The latter assist her and enlarge the nest by building channels or passageways, cleaning the cells, finding food for the larvae and airing out the nest. By the end of the season, 3,000 wasps may occupy an 8-inch to 12-inch (20-cm to 30-cm) diameter nest. The sexual individuals mate in their nest in the fall. The young fertilized queens, the only survivors of the colony, then hibernate either in a barn or somewhere in a crevice.

If you touch it, you will be stung

A Japanese proverb states: "Wasps only sting crying faces" (meaning adversity touches the saddest). However, wasps do warn us: their distinct markings (yellow-and-black stripes) remind predators and everyone else that their stings can hurt. They are painful and can be quite dangerous in large numbers, particularly if the person stung is allergic to the venom. It is a known fact that more people die of wasp stings than of venomous snake bites.

ACTUAL SIZE

Vespula vulgaris

(Linnaeus, 1758)

Size: Worker: 11–14 mm

Male: 13–17 mm

Queen: 16–19 mm

XYLOCOPA VIOLACEA

European Carpenter Bee

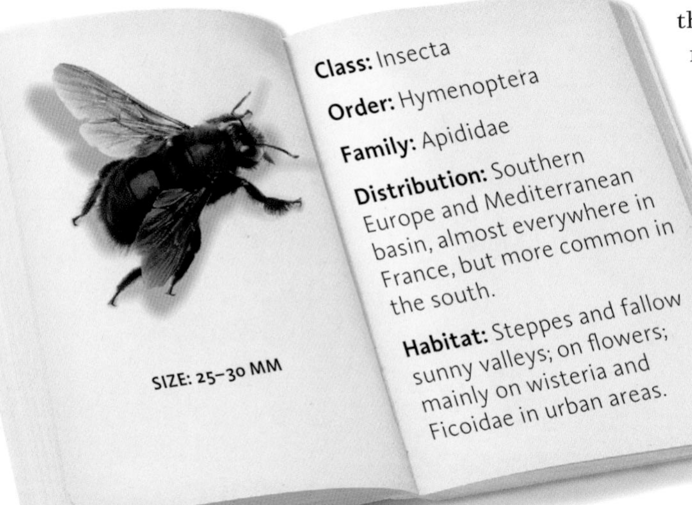

Origin of the name

The word *Xylocopa* refers to the drilling behavior (from *xulos*, or wood, in Greek) of these insects, and *violacea* is a reference to the metallic blue-violet sheen of their bodies. This species is presently known as the Carpenter Bee since it builds its nest inside wood.

Impressive but harmless

Despite its large size, the Carpenter Bee does not attack humans. Nevertheless, males, recognizable by the yellow rings at the tip of their antennae, have no stinger and females only rarely sting. The bumblebee-like flight of the Carpenter Bee is fast and vigorous as it circles flowers rich in nectar, especially those of wisteria and the Ficoidae family. This bee is found almost throughout France, but is more abundant in the south, where similar species of Carpenter Bees (notably *X. valga*) also reside. It seems that climate warming has contributed to its northward expansion.

Nymphs in their loge

Carpenter bees

This species comprises the largest and stockiest solitary bees in Europe. Both females and males hibernate in barns or hollow trees. They wake up in the spring and mate. To establish her nest, the female selects perfectly dry wood structures, be they posts, dead tree branches or support beams. Using her extremely strong mandibles (photo at bottom left), she begins by drilling a horizontal tunnel opening (many natural openings are often filled). After that comes a vertical tunnel about 0.5 inches (1.5 cm) in diameter and 6 to 12 inches (15 to 30 cm) long, with but a single exit. At the end of this 10 to 15 cells are constructed, separated by walls of compacted sawdust glued in place with saliva. Each cell contains a few grams of pollen and a larva. After turning into nymphs at the end of August or in September, the young adults emerge using the unique opening of the nest as guide.

A keen sense of orientation

Aided by her very long (0.4 inch/1 cm) tongue to reach deep into corollas, the female Carpenter Bee tirelessly gathers nectar and pollen from nearby flowers. If the corolla is too deep, like those of narcissus flowers, for example, she makes an incision at the base using her sharp mouthparts to be able to reach the nectar. If there are not enough flowers in the vicinity, thanks to a remarkable sense of orientation, the Carpenter Bee will travel more than 3 miles (5 km) from her nest in search of nectar.

The mandibles of the European Carpenter Bee – a veritable carpenter's gouge!

LOYAL TO THEIR HOME BASE

Provided there are enough flowers in the area (they are especially attracted to wisterias) Carpenter Bees are often found in urban locations. They also tend to be stay-at-home insects, since each year new young females tend to build their nests close to where they were born. For example, a building belonging to Cornell University in New York State has housed the same population of Carpenter Bees (*X. virginica*) since 1930. The population was still thriving in 2006!

Nest entrance

Xylocopa violacea

(Linnaeus, 1758)

Size:

25–30 mm

Zygaena filipendula

Six-spot Burnet Moth

Origin of the name

When Linnaeus mistakenly first described this moth, thinking that it developed on the Filipendula Spiraea (Queen of the Meadow plant – *Filipendula vulgaris*), he therefore gave it the species name. In fact, the caterpillar of this butterfly feeds on trefoils and deervetches (*Lotus* species). The genus name *Zygaena*, coined by Fabricius (from the Greek, *zugon*, for "yoke," and *ainos*, for "terrible, frightful,") refers, in an exaggerated manner, to the thick, forward-thrusting antennae. These moths were once called "sphinxlike sheep"!

A diurnal nighttime moth

The Six-spot Burnet Moth, the most common of its genus, has in general six very distinct red spotted markings. Although the rooflike positioning of its wings and antennae at rest are characteristic of heterocerate (nighttime) moths, it is also active during the day. These moths inhabit very diverse biotopes.

Flying poison

Six-spot Burnet Moths are a bit skittish. Their very contrasting and characteristic markings remind predators that they are toxic. This is called aposematism (visually warning coloration).

When first threatened, the Six-spot Burnet Moth secretes bad-tasting substances from glands located at the base of its "horn" and its tarsus. These substances consist of volatile, acrid and corrosive compounds known as pyrazines and are sufficiently noxious to keep predators at bay. Should the attacker persist, secretions of cyanide will ooze from its leg and thorax articulations. Though produced in small quantities, this incredibly toxic poison is enough to further detract predators. It originates from the host plants consumed by the caterpillars – including diverse *Lotus* plants – which contain a substantial amount of cyanide. Despite this impressive chemical arsenal, the Six-spot Burnet Moth does fall prey to certain birds and to a small number of spiders.

Inspiring Mimics

Six-spot Burnet Moths have a foundation of red spots (or white ones) on a blue or metallic green background. Their "success" at repelling predators appears to have "inspired" other insects to mimic them and thereby benefit from similar protective coloration – without themselves being toxic. In butterflies, for example, the Amata (family Arctiidae) have the similarly shaped wings, spots and stomach belting. There are also insects of other orders that imitate these moths; for example, members of the Pentatomidae as well as coleopterans like the Cerambycidae and Meloidae.

Zebra-striped caterpillars

The typical caterpillars (yellow with pointed black stripes and a velvety texture) feed on plants in the *Lotus* genus. They hibernate and become nymphs the following spring by weaving a brilliant, yellow parchment-like cocoon on a blade of grass. Adults emerge in July.

The chrysa

Egg laying

The Burnet caterpill

ACTUAL SIZE
N.° Zygaena filipendula
(Linnaeus, 1758)
Wingspan:
32–42 mm

Italicized words are referenced in this glossary.

ametabolic (adj.) Term that applies to an insect whose development does not go through *metamorphosis.*

apode (adj.) An organism that does not have feet, such as the larvae of several insects.

apterous (adj.) Having no wings.

article (n.) In arthropods, it is the part of one appendage that articulates on another.

cephalothorax (n.) Anterior region of some arthropods (spiders, for example) arising from the fusion of the head and the thorax.

chelicerae (n.) Appendage located in the anterior body region of chelicerate arthropods. They are transformed into venomous fangs in spiders.

chitin (n.) Main constituent of the arthropod cuticle (as well as some mushroom tissues). Chitin is a macromolecule, a nitrogenous polysaccharide (like cellulose in plants).

chrysalis (n.) The *pupa* of a butterfly residing in the developmental stage between caterpillar and adult. In nocturnal butterflies, the chrysalis is encased in a silk cocoon.

clone (n.) A group of genetically identical individuals conceived from a single unique individual.

communal (n. & adj.) Term used to describe an animal species that lives off another animal's leftovers, without harming it.

cosmopolitan (adj.) Term used to describe a species widespread throughout the world.

cryptic (adj.) That which hides an animal in its natural environment.

cuticle (n.) External, and impermeable rigid layer of arthropod skin composed mostly of *chitin.*

diapause (n.) Period of slowing in the developmental activity in certain insects; can be seasonal or last for several years.

elytron (n.) (plural: elytra) The soft and leathery, or tough and hard nonfunctional forewings of beetles; they serve to protect the delicate, membranous hindwings (flight wings), as well as the internal organs inside the abdomen.

entomophagous (n.) The process of feeding on insects.

exuvia (n.) (plural: exuviae) Skins left behind after an insect has *molted.*

femur (n.) One of the segments of an insect's leg, located in a median position.

gonopod (n.) Reproductive male organ in many arthropods.

hemelytron (n.) (plural: hemelytra) One of the forewings of a true bug, having a hard, thick basal portion and a thinner membranous apex.

hemimetabolic (adj.) Refers to insects undergoing metamorphosis that lack a pupal stage. The young do not resemble the adults.

hemolymph (n.) The circulating liquid in arthropods carrying nutrients to body tissues and removing metabolic waste. It is the "blood" of arthropods.

heterometabolic (adj.) Refers to insects undergoing incomplete metamorphosis in which the nymph is essentially an adult (imago) and there is no pupal stage.

homometabolic (adj.) Refers to insects undergoing complete metamorphosis. In this type of development, the larval stage is markedly different from the adult.

honeydew (n.) Sugary product resulting from some insects' (notably aphids) digestion of plant sap matter.

imago (n.) Adult form of insects fully developed and ready to reproduce.

labrum (n.) (plural: labria) Upper "lip" of insect mouthparts.

larva (n.) (plural: larvae) Stage of insect development between the egg and the adult. Larvae undergo a series of molts and the stage between each molt is an *instar.* The number of instars varies according to species, gender and environmental conditions. Larvae increase in size with each molt until they reach adulthood.

mandible (n.) Mouthpart of arthropods ahead of jaws. They generally serve to grasp, bite, cut, chew grind, or pierce plant and animal food. They may be transformed into weapons either for defense, offense (soldier ants), or fused to maxillae to form a stinging apparatus (mosquitoes and gadflies).

markings (pl. n.) The ensemble of coloration and shapes on the body of an animal.

metamorphosis (n.) (plural: metamorphoses) Bodily transformation of insects during development from eggs to adulthood. Different types are known: the complete type, a radical change from larva to adult (*imago*); and incomplete, consisting of successive molts during which the wings and reproductive organs are progressively developed.

molt (n.) In arthropods, this refers to a changing of the "skin," allowing growth. During the molt, the insect rids itself completely of its exoskeleton (hence known as the *exuviae*), after which a new cuticle is laid down and hardens into a new exoskeleton. Insects can go through two to several molts in one generation.

nymph (n.) Aquatic or terrestrial larva of insects that develop by complete metamorphosis. The nymph is known as a *chrysalis* in lepidopterans (butterflies) and as a *pupa* in dipterans (flies).

ocellus (n.) (plural: ocelli)
1. Simple eye in many arthropods, such as insects, spiders, etc., perceiving variations in light intensity (unlike compound eyes that perceive images).
2. Round eyelike wing decorations in some insects (notably in butterflies). Ocelli, mimicking the eyes of much larger animals, can serve to deter predators.

ovipositor (n.) Elongated egg-laying organ located at the extremity of the abdomen in some female insects (cicadas, grasshoppers, parasitic hymenopterans), it can pierce animal and plant tissues, or drill into soil to deposit the eggs.

palp (n.) A tiny anterior appendage in arthropods that have sensory structures involved in taste, smell and touch. Usually in the form of two pairs: the maxillary palps and the labial palps.

parasite (n.) (adj.: parasitic) An organism that lives exclusively in or on another organism (its host) and feeds off it.

parasitoid (n. & adj.) Insects where the female lays her eggs in or on the body of another organism of a different species. The aphids are an example of this. Eventually, the host will be killed.

parthenogenesis (n.) An asexual mode of reproduction where young develop from unfertilized eggs. In many insects, such eggs develop into females genetically identical to their mother, although in ants, bees and wasps, they always develop into males.

pedipalp (n.) In arachnids, a pair of leg-like appendages located on either side of the chelicerae that are used as fingers to manipulate prey. Mature male spiders have special sperm-storing organs on the tips of their pedipalps.

pheromones (n.) Chemicals released outside the body in very small amounts to influence the behavior of individuals of the same species. Pheromones are used to attract mates, mark trails or send alarm signals.

phytophagous (adj.) Plant-eating.

polyphagous (adj.) Feeds on a variety of plant matter.

pronotum (n.) Dorsal and anterior portion of an insect's thorax forming in some (coleopterans, bugs, cockroaches) a rigid armor occasionally entirely covering the head.

propolis (n.) A resinous mixture that honeybees collect from tree buds, sap flows or other plant sources. It is used to build and repair the hive.

pupa (n.) (plural: pupae) Name given to the nymph or final larval stage of dipteran insects (flies, mosquitoes). Pupae are sometimes wrapped in a protective material such as a paparium or a cocoon.

rostrum (n.) Anterior-most portion of the head in some insects; has mouthparts at its extremity which it can pierce and suck with (aphids and bugs) or with which it can chew (weevils).

scutellum (n.) Triangular (in coleopterans) rigid plate on the posterior face of the thorax of some insects, at the level of the hindwings. It can be round in bugs.

spermatophore (n.) In many arthropods (spiders, grasshoppers, some millipedes) a capsule or packet of sperm released by the male close to the female's reproductive organs or on the ground.

stabilizer (n.) Steadying flight structure/organ in dipteran insects. These are located on the hindwings.

stinger (n.) Hollow, deep and pointed venomous organ, located at the extremity of the abdomen in hymenoteran insects (bees, wasps).

tarsus (n.) (plural: tarso) In hexapods, arachnids, millipedes and centipedes, the outermost segment(s) of the leg, attached to the *tibia* in insects and to the metatarsus in arachnids. The tarsus can be made of one or up to five segments.

telson (n.) In arthropods, the terminal segment of abdomen. In some arthropods, such as scorpions, the final body segment, curved up in a stinging tip.

tibia (n.) (plural: tibiae) One of the leg segments located between the femur and the tarsus.

tracheobranchiae (pl. n.) In aquatic insects, respiratory organs able to capture the oxygen dissolved in water. The tracheobranchiae are body extensions in the form of tiny tubes.

English names are in roman type and scientific names are italicized. The names in **boldface** indicate species illustrated in full plates in this book. Page numbers followed by (T) refer to species illustrated by thumbprint only.

Written Text
With the Collaboration of
Patrice Leraut

Managing Editor
Catherine Delprat

Editors
Thierry Olivaux, assisted by **Alice Renée**

Editing and Proofreading
Julie Lecomte and **Aurélia Licatesi**

Design Conception and Artistic Direction
Emmanuel Chaspoul, assisted by **Cynthia Savage**

Layout
Natacha Marmouget

Cover
Véronique Laporte

Production
Annie Botrel

Photoengraving
Nord Compo

English Translation
Margaret and **Klaus Brasch**

Photo Credits

Photographs: © Sonia Dourlot, except for:

Bernard Chaubet: p. 26: caterpillar; p. 26: chrysalis; p. 44: caterpillar; p. 48: wingless female; p. 58: larva; p. 62: mating; p. 62: larva; p. 64: eggs; p. 78: on a flower; p. 86: predation; p. 110: larva; p. 122: mother caring for her eggs; p.130: caterpillar; p. 130: chrysalis; p. 136: male chirping; p. 168: on a Marsh Gentian flower; p. 172: larvae; p. 172: infested bee; p. 192: young larvae; p. 196: caterpillar with osmetrium (in box); p. 234: mating; p. 242: on fruit.

Larousse Archives: p. 7 (bottom). **Denis Bourgeois**: p. 34: mating; p. 56: in a group; p. 172: mating; p. 218: larva. **Simon Cavaillés:** p. 156: female. **Glenn Dubois;** p. 190: cavity in an old tree. **Pascal Dubois;** p. 76: male. **Pierre Duhem:** p. 242: on a flower. **Parasitoid Insect Team/Rennes University 1:** p. 214: Sirex larva. **Alexandre François:** p. 184: larva. **Bruno Lavoué:** p. 68: larva; p. 78: defensive stance. **Bernard Le Garff:** p. 166: caterpillar; p. 188 female; p. 216: coupling; p. 238: male; **LNPV Montpelier 9:** p. 212: larva. **OPIE/Rémi Coutin:** p. 204: larva. **Olivier Ploton/Larousse Archives:** p. 8, p. 13 (bottom), p. 15 (bottom), p. 16 (bottom), p. 17 (upper). **Christian Pourre:** p. 144: female. **Jean-Yves Robert:** p. 216: larva. **Heinz Rothacher:** p. 162: male combat; p. 166: caterpillar. **Sylvain Sanchez:** p. 120: mother and her young. **Christian Vanderbergh:** p. 34: larva; p. 246: nymph.

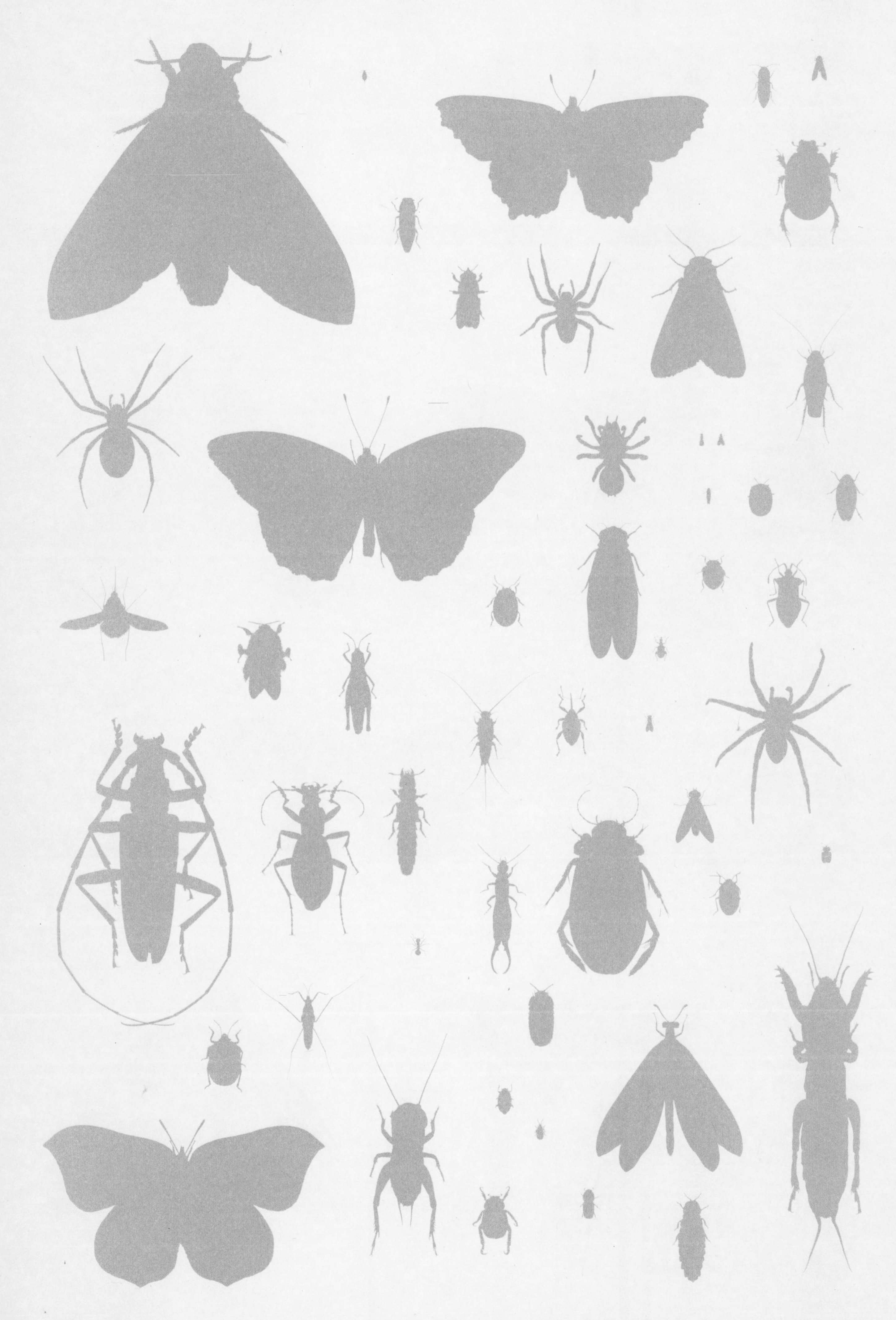